Design of Wastewater and Stormwater Pumping Stations

Manual of Practice FD-4

Prepared by **Task Force on Design of Wastewater and Stormwater Pumping Stations**
Terry L. Krause, *Chairman*

Diego Alaimo
Richard L. Anderton
Robert J. Andryszak
Peter J. Barthuly
Henry Benjes, Jr.
Aivars Bergs
Douglas G. Brinkman
Gerald T. Carey
Steven W. Collins
Tracy O. Dubcak
Joseph Gemin
Frank M. Gunby, Jr.
Paul J. Higgins
Barton G. Jones
Madhusudan C. Joshi
Louis W. Klinefelter
J. Robert Krebs
James D. Linsley
Chung-Lyu Liu
Michael L. Middleton
Darrell Morrison
Roy Parsons
Manu A. Patel
Calvin C. Patterson
Charles M. Pekkala
Thomas A. Ridgik
Paul R. Rom
Thomas M. Simpson
Adam H. Slutsky
Peter M. Smith
Wilson F. So
Marc S. Solomon
Srinivasan Sundaramoorthy
Thomas G. Thompson
Rufino A. Vigilia, Jr.
Thomas M. Walski
Thomas L. Weber
Glenn R. Wentink
H. Leroy Whiteley, Jr.

Under the Direction of the
Facilities Development Subcommittee of the Technical Practice Committee

1993

Water Environment Federation
601 Wythe Street
Alexandria, VA 22314–1994 USA

Library of Congress Cataloguing-in-Publication Data

Design of wastewater and stormwater pumping stations/prepared by Task Force on Design of Wastewater and Stormwater Pumping Stations of the Water Environment Federation.

p. cm.—(Manual of practice; FD-4)

Includes index.

ISBN 1-881369-36-6: $35.00 (est.)

1. Pumping stations—Design and construction. 2. Sewerage—Design and construction. I. Water Environment Federation. Task Force on Design of Wastewater and Stormwater Pumping Stations. II. Water Environment Federation. Facilities Development Subcommittee. III. Series: Manual of practice. FD; no. 4.

TD725.D47 1993

628'.29—dc20 93-6136

CIP

Alexandria, VA 22314–1994 USA

Library of Congress Catalog Card No. 93-6136

ISBN 1-881369-36-6

Printed by Imperial Printing Company **1993**

Manuals of Practice for Water Pollution Control

The Water Environment Federation (WEF) Technical Practice Committee (formerly the Committee on Sewage and Industrial Wastes Practice of the Federation of Sewage and Industrial Wastes Association) was created by the Federation Board of Control on October 11, 1941. The primary function of the committee is to originate and produce, through appropriate subcommittees, special publications dealing with technical aspects of the broad interests of the Federation. These manuals are intended to provide background information through a review of technical practices and detailed procedures that research and experience have shown to be functional and practical.

IMPORTANT NOTICE

The contents of this publication are for general information only and are not intended to be a standard of the Water Environment Federation (WEF).

No reference made in this publication to any specific method, product, process, or service constitutes or implies an endorsement, recommendation, or warranty thereof by the Federation.

The Federation makes no representation or warranty of any kind, whether expressed or implied, concerning the accuracy, product, or process discussed in this publication and assumes no liability.

Anyone using this information assumes all liability arising from such use, including but not limited to infringement of any patent or patents.

Preface

This manual of practice presents contemporary design considerations for wastewater and stormwater pumping stations. The manual, which replaces the previous pumping station design manual published in 1981 by the Water Environment Federation (then Water Pollution Control Federation), covers station capacity requirements and configuration, the selection and design of pumping equipment, piping system layout and hydraulic considerations, electrical and instrumentation systems design, and other station appurtenances.

The manual is intended for the practicing design professional. Accordingly, it focuses on current design practice for those individuals that have a general knowledge of the basic principles of pumping station design. Considerations discussed are intended to cover the design of both small and large pumping stations and highlight the differences between wastewater and stormwater stations.

The contents of this manual represent the collected background and experience of more than 40 professionals active in the design of wastewater and stormwater pumping stations. Material presented is intended to complement other Water Environment Federation manuals of practice and published design guidelines and textbooks.

This manual was produced under the direction of Terry L. Krause, *Chairman.* The principal authors are

Richard L. Anderton
Steven W. Collins
Frank M. Gunby, Jr.
J. Robert Krebs
James D. Linsley
Thomas A. Ridgik
Wilson F. So
Carlos E. Vargas

In addition to Task Force and Technical Practice Committee Control Group members, additional information and review was provided by Joseph A. Guagno, Bill Peterschmidt, Bill Phillips, and Ted Stolinski.

Authors' and reviewers' efforts were supported by the following organizations:

Baltimore Division of Wastewater Engineering
Black & Veatch
Boyle Engineering
Buchart Horn
C.C. Johnson & Malhotra
CH2M Hill
EMA
EnecoTech
FHC, Inc.
Greeley and Hansen
Hayden-Wegman
HDR Engineering
HNTB
Horner & Shifrin
KCI Technologies
Krebs Consulting Services
Lewis and Zimmerman Associates
Metcalf & Eddy
Potter & Parsons
Proctor & Redfern LTD
PSC Engineers & Consultants
Roy F. Weston
RV Anderson Associates
So & Associates Engineers Inc.
Strand Associates
Weston & Sampson
Whitman & Howard
Whitman Requardt & Associates
Williams and Wentink

Federation technical staff project management was provided by Greg White; technical editorial assistance was provided by Bill Wolle and Lorna E. Johnson.

Abstract

This manual presents current design practice for wastewater and stormwater pumping stations. It is intended to be a comprehensive and useful source of practical design information for the practicing design professional. Chapter 1 describes the purpose and scope of the manual and guidance to its use. Chapter 2 reviews methodology to establish station capacity. Chapter 3 presents station layout and configuration considerations. Chapters 4 and 5 review pumping and piping system design, respectively. Chapter 6 discusses station electrical design and Chapter 7 reviews instrumentation and control system design. Finally, Chapter 8 discusses design considerations for typical station appurtenances.

The Water Environment Federation is a nonprofit, educational organization composed of member and affiliated associations throughout the world. Since 1928, WEF has represented water quality specialists, including biologists, bacteriologists, local and national government officials, treatment plant operators, laboratory technicians, chemists, industrial technologists, students, academics, equipment manufacturers/distributors, and civil, design, and environmental engineers.

For information on membership, publications, and conferences, contact

Water Environment Federation
601 Wythe Street
Alexandria, VA 22314–1994 USA
(703) 684-2400

Contents

List of Tables

List of Figures

Chapter 1
Introduction

OVERVIEW

Wastewater and stormwater collection and conveyance systems are extensive networks of interconnected piping. Because of the expansiveness and topography of these systems, wastewater or stormwater flows cannot always be conveyed by gravity. As a result, pumping stations are needed to lift the flow over high points in the systems or to treatment facilities. Because these stations play a vital role in the collection system, design engineers must produce plans for well-engineered pumping stations that operate reliably and efficiently.

Wastewater or stormwater pumping stations range from simple facilities of low capacity in remote areas to complex facilities capable of pumping billions of litres (gallons) of wastewater or stormwater each day. Many times, especially at the influent end of a treatment facility, the pumping station is combined with other preliminary treatment systems. As a result, the design of these facilities is a complex undertaking that requires an understanding of all the design engineering disciplines. The design engineer should use all of the technical disciplines early in the design process to ensure a well-coordinated and functional pumping station.

This manual has been prepared to assist the practicing engineer in the design of wastewater and stormwater pumping stations and is for the experienced designer rather than the novice. While there are many similarities between the design of wastewater and stormwater pumping stations, there are differences. For example, while there is similarity between wastewater and stormwater pumps, pipes, valves, and controls, there are differences in station hydraulics and operational considerations such as the intermittent operation of a stormwater pumping station.

GUIDE TO THE MANUAL

The content and organization of the manual is intended to assist in the planning and design of both wastewater and stormwater pumping stations. The manual is divided into eight chapters and should be used in conjunction with other Water Environment Federation manuals that deal with the design of collection and treatment facilities.

Chapter 2, "Station Capacity," covers the methods of determining the need for a pumping station and the station's capacity. Further discussions relative to capacity are covered in *Design and Construction of Sanitary and Storm Sewers* (1969).

Chapter 3, "Station Configuration and Design," presents information on facility location, types, aesthetic and safety considerations, and architectural and structural design. Wet well configuration and storage considerations also are addressed.

Chapter 4, "Pumping Systems," reviews pumping system definitions and describes types of pumps and pump drives typically used. The chapter also presents pumping system design concerns such as pump curve preparation and pump selection criteria.

Chapter 5, "Piping Systems," reviews fluid flow fundamentals, water hammer and surge considerations, and the design of station piping systems, including piping arrangements, piping materials, valves, and flow-metering considerations. Additional design information covering station components can be found in *Design of Municipal Wastewater Treatment Plants* (1992).

Chapter 6, "Electrical Design," describes layout of a pumping station's electrical system, including standby power considerations. Additional design information can be found in *Design of Municipal Wastewater Treatment Plants* (1992) and *Energy Conservation in the Design and Operation of Wastewater Treatment Facilities* (1982).

Chapter 7, "Instrumentation and Control Systems," presents information about metering elements, gauges, system alarms, controls, and telemetry systems. Additional design information can be found in *Instrumentation in Wastewater Treatment Plants* (1993).

Chapter 8, "Appurtenances," discusses design considerations for typical appurtenances found in pumping stations, including screening, odor and noise control, and building services such as heating and ventilation, lighting, water supply, drainage, utilities, hoists, and safety features. More complete coverage of odor control is contained in *Odor Control for Wastewater Facilities* (1979). Additional design information on most station appurtenances can be found in *Design of Municipal Wastewater Treatment Plants* (1992).

For specific information on pumping station operation and maintenance considerations, refer to *Operation of Municipal Wastewater Treatment Plants* (1990).

REFERENCES

Design and Construction of Sanitary and Storm Sewers (1969). Manual of Practice No. 9, Water Pollut. Control Fed., Washington, D.C.; Manual of Engineering Practice No. 37, Am. Soc. Civ. Eng., New York, N.Y.

Design of Municipal Wastewater Treatment Plants (1992). Manual of Practice No. 8, Water Environ. Fed., Alexandria, Va.; Manual of Engineering Practice No. 76, Am. Soc. Civ. Eng., New York, N.Y.

Energy Conservation in the Design and Operation of Wastewater Treatment Facilities (1982). Manual of Practice No. FD-2, Water Pollut. Control Fed., Washington, D.C.

Instrumentation in Wastewater Treatment Plants (1993). Manual of Practice No. 21, Water Environ. Fed., Alexandria, Va.

Odor Control for Wastewater Facilities (1979). Manual of Practice No. 22, Water Pollut. Control Fed., Washington, D.C.

Operation of Municipal Wastewater Treatment Plants (1990). Manual of Practice No. 11, Water Pollut. Control Fed., Alexandria, Va.

Chapter 2
Station Capacity

SANITARY WASTEWATER FLOW

Sanitary wastewater consists primarily of used portions of public or private potable water supplies that are discharged to sewers through the plumbing systems of residential buildings. They may also include portions of a public or private water supply discharged through the plumbing systems of industrial and commercial establishments. Sanitary wastewater also contains an inflow and infiltration (I/I) component. The ratio between residential, industrial, and commercial and I/I components will vary from one sanitary waste collection system to another. Before final discharge, these flows normally are collected and carried to a treatment facility where various pollutants are reduced or removed.

To provide economical and effective treatment for sanitary wastewaters, their volume should be kept to a minimum and not combined and diluted with larger volume stormwater flows, infiltration, or inflow. Inflow, such as storm runoff from roof, yard, and foundation drains, typically is collected separately from sanitary wastewater and conveyed to a disposal point by a separate sewer or drainage system designed for this purpose.

ENGINEERING REPORT. Before the design and construction of a sanitary wastewater collection and pumping station, an engineering report detailing the basis of design should be prepared. This report should analyze associated planning studies, drainage areas, population growth projections, land development, quantity of available water, projected quantity of wastewater flows—including all significant wastewater contributors—I/I, and the quality of influent water. The engineering report often includes the design schedule for the project and estimated costs of alternatives.

Based on the results of these preliminary studies, the engineering report should be expanded, if applicable, to include the pumping station basis of design. The basis of design should include the range of expected flows for which the facility will be designed, the number and capacity of pumps, the design life, and the design criteria for each engineering discipline (that is, civil, structural, architectural, mechanical, electrical, and instrumentation and control) involved. The basis of design should include an estimate of the construction cost, a preliminary assessment of the type of equipment, need for screening or other pretreatment, reliability standards (standby pumps and backup power supplies), need for station amenities (such as restrooms and offices), general facility layout, equipment access, control system, monitoring needs, overall pump station aesthetics, and odor control considerations (Sanks *et al.*, 1991).

This chapter addresses development of the facility design life and determination of facility capacity. Subsequent chapters address the facility basis of design.

Design Life. There is a maximum hydraulic or functional capacity beyond which a wastewater system will not operate satisfactorily. The time span, a period ranging from 10 to 30 years from the date of the original design to an estimated future date when the capacity of the system will be reached, is referred to as the "design life."

The system's design life should be determined before the start of any engineering design. An analysis of the design life should take into account the following:

- Anticipated obsolescence and wear and tear on the mechanical equipment,
- The possibility of needing to expand the system in the future,
- The ease or difficulty of implementing such an expansion, and
- The efficiency of the system's performance during its initial years when it may not yet be operating to full capacity.

In choosing a design life, undue financial burden on the system's present users should be avoided. A longer design life likely will require that larger facilities be constructed, resulting in higher construction and perhaps operation and maintenance costs. Selecting too short a design life could result in additional facility expansion costs sooner than desired. Careful selection of the design life is important to ensure that the system's cost be shared equitably by existing and future users.

In summary, selecting a reasonable design life should reflect the ability of the design to accommodate the flexibility to meet future conditions through planned modification options (impeller and motor changes) or expansion of standard elements (pumps).

Population Growth Projection. The quantity of sanitary wastewater generated by a community depends largely on the population and the per capita wastewater contribution. Therefore, an accurate population projection is essential to implementing a well-planned collection and treatment system. A community's population growth and distribution may be affected by local economics, employment opportunities, commuting distances for workers, socioeconomic factors, zoning, availability of community facilities and services, and other community amenities. Factors not evident at the time of the study may emerge to cause a major shift in population growth, for example, the discovery of new natural resources in the vicinity, availability of new or expanded water supplies, or an industry's decision to move to or from the community.

Historical population information can be obtained from various sources. The U.S. government decennially has conducted a census since 1790. The census provides valuable demographic information on population, such as socioeconomic characteristics, housing facts, and population per household.

Additional data for the interim period may be obtained from local surveys for schools, building permit records, tax records, chamber of commerce studies, local public utility records, and local and regional planning commissions. Careful analysis of census data may reveal past population trends.

There are various methods of projecting population trends, depending on the amount and types of historic data available. These are determined by the following factors:

- Arithmetic (changes population by a fixed constant quantity);
- Geometric (changes population in proportion to population size, by a fixed ratio);
- Cohort survival (determines population by age groups and sex);
- Graphical (comparison with the growth of a similar but larger community);
- Apportionment (establishes population based on projections for larger dimensions, such as county or state); and
- A combination of the above.

Because some of these methods may be complicated, especially when used in combination with one another, the desired degree of accuracy should be considered when projection techniques are selected. The engineer should check with the local planning agency first, because in many cases population projections may already have been developed in sufficient detail. Population projections should be reviewed with the local planning agency and the project owner before proceeding with the study.

Details of population projection methods are discussed in many textbooks and references (McJunkin, 1964; McLean, 1952; and a U.N. publication, Population Projections). Most graphical or mathematical projections are based on extending past growth trends. These methods are more applicable to communities with historically stable growth. For young and fast-growing communities, or communities with high growth potential, a meaningful projection requires good engineering judgment based on past information coupled with consideration of all related environmental determinants. Similarly, consideration should be given to the possibility of reduced growth rates in older or more mature communities at or near growth saturation.

Population projections can also be obtained from previous studies prepared for the community. Previous U.S. Environmental Protection Agency (U.S. EPA) funded construction grant studies contain population projections for the community or for larger basin areas. These projections may require further breakdown for meaningful use on a single project covering a smaller area.

Land Development. Local and regional land-use planning, development, urban renewal projects, and zoning regulations will influence future population growth and distribution, which in turn will affect the quantity of wastewater generated by a community. In many urban areas, burgeoning construction of multidwelling apartments, condominiums, and town homes have led to increased population densities and wastewater quantities.

Many housing units are equipped with appliances (such as dishwasher, garbage disposal unit, or clothes washer) that increase water consumption and the amount of sanitary wastewater generated. However, rising costs of water in some communities have tended to suppress water use, hence reducing wastewater flow.

Quantity of Available Water. The quantity of wastewater flow generated is closely related to the quantity of available water and its consumption rate. In addition, the public's general tendency to use water will vary depending on weather conditions, metered or unmetered water supplies, water rates, local conservation efforts, availability of public sewers, and the economic condition of the community.

Because wastewater consists primarily of used water, which is often metered, per capita water consumption can be established. From this information, it is possible to estimate the per capita quantity of wastewater reaching the sewer. Records of water usage (daily, monthly, and seasonally) typically are available from the water utility that provides water service to the study area. National and regional water usage rates may also be available. Annual fluctuations in water demands and, subsequently, wastewater flows also play an important role in design of sewerage systems, hence in the required capacity of pumping stations.

Quantity of Wastewater Flow. The quantity of wastewater flow varies continuously in a wastewater collection system. There will be a wide fluctuation of flow throughout the day in any system where sanitary wastewater is a major contributor to the total flow. If wastewater facilities are not designed to accommodate these fluctuations, they can and will cause operational problems in collection and pumping systems and in treatment facilities.

For design purposes, it is necessary to estimate future wastewater flows based on existing conditions. State and municipal regulatory agencies often provide average per capita domestic wastewater design flows or curves of flow versus population to evaluate required pumping station capacity. If no such guidance is available and the system has no flow records, or if a complex sewer network makes actual measurement of flows impractical, design flows may be determined by either estimating and totaling the flows from the various contributors or by calculating total flow as a percentage of the total available water supply. In either case, careful study and judgment are needed to project wastewater flow because average flows have been shown to vary between 190 and 1 000 L/cap·d (50 and 265 gpd/cap) (Gravity, 1982).

Engineers typically are interested in the average daily flow, maximum and minimum daily flows, and peak hourly flow in a sanitary sewer collection system for a variety of different design purposes. For pumping stations, the capacity of required pumping equipment and the pumping station wet well volume will depend on either the peak hourly or maximum daily flow, or a combination of the two (that is, the peak hour of the maximum day), whichever is greater. Peak hourly flow will determine the maximum required pumping capacity of the station. However, minimum daily flows also are important in pumping station design, not only to assess maximum residence time in wet wells, but also to determine required turndown capacity of the pumping equipment. Turndown capacity is the ratio of maximum to minimum pumping capacity of the pumping station (for example, a required turndown capacity of 2:1 would indicate that the minimum pumping capacity at which the pumping station will have to operate effectively is 50% that of the maximum capacity).

Additional Contributors. In addition to domestic wastewater flows, contributions may come from commercial, industrial, institutional, and I/I sources.

COMMERCIAL. Wastewater flows from commercial districts often are estimated in m^3/ha·d (gpd/ac), based on information from existing similar establishments, water consumption, and comparative data from other cities. Flow allowances for sewer design can vary from 42 000 to more than 1 500 000 L/ha·d (4 500 to 160 000 gpd/ac) in different cities and depend on the type of commercial activity (Gravity, 1982). Generous allowances for many commercial establishments may be justified because of the possibilities of future expansions and improvements.

INDUSTRIAL. Two basic types of wastewater typically are generated by industries in addition to the sanitary flow originating from employees: process wastewater and spent cooling water. For existing industries, the quantity of process wastewater may be obtained through a field survey and flow monitoring or by analysis of water consumption records. For future industrial facilities, wastewater flow may be projected by studying the processes to be included in the new facility, the water consumption rate for each process, time of day and duration of process operation, types and numbers of industrial buildings, estimated number of employees, and comparative data from similar industries.

Spent cooling water to be handled by sanitary systems normally comes from water-cooled air-conditioning chiller units. The average water demand for once-through, water-cooled air-conditioning equipment ranges from 0.10 to 0.14×10^{-3} L/kg·s (1.5 to 2 gpm/ton) of refrigeration. Water for multiple-pass cooling systems and for major industrial processes generally is treated and handled by separate systems. A small amount of water generally is wasted or "blown down" to the sewer to maintain system stability. Consideration must be given to this water because, while it may be small in volume, it may contain a high proportion of certain pollutants such as water-conditioning additives and total dissolved solids.

INSTITUTIONAL. The quantity of institutional wastewater may be established based on such items as the population served, the number of buildings, the number of units of equipment, hygiene controls, and individual habits. The design flow can be determined by field survey of the existing facilities or by analyzing comparative data from similar institutions.

Table 2.1 summarizes typical wastewater flows generated from both residential and nonresidential (commercial and industrial), institutional, and recreational (seasonal) establishments. Flows from these establishments can be added to determine the required pumping station capacity needed to serve an existing or new community.

INFILTRATION/INFLOW. When determining wastewater flows, allowance also should be made for miscellaneous leakages to the sanitary systems. These include infiltration through the sewer pipes and pipe joints, leakage through manholes and various sanitary facilities, and a certain amount of stormwater that inevitably finds its way into the sanitary collection system. However, in a properly designed, installed, and well-operated and maintained collection system, the I/I rate should not exceed 18 520 (mL/m·d)/m (200 gpd/in. dia/mile) for any section of the collection system (Domestic Wastewater, 1991). Newer collection systems, constructed after 1970, should achieve I/I rates of between 13 900 and 18 520 (mL/m·d)/m (150 and 200 gpd/in. dia/mile), while older systems with less efficient joint construction may experience leakage in excess of 27 780 (mL/m·d)/m (300 gpd/in. dia/mile).

STORMWATER FLOWS

The quantity of flow reaching a stormwater pumping station is determined by hydrologic analysis of the regional rainfall events; the characteristics of the watershed; and the hydraulics of the collection, treatment, and discharge system (Chow, 1964; Chow *et al.*, 1988; Linsley *et al.*, 1958; and Williams, 1950). In many instances, storm drainage systems are designed solely on the basis of peak runoff values, thus neglecting flow variations with time and routing of the flow through the watershed, the collection and treatment systems, and any storage facilities along the flow route. This method is an adequate but conservative means of sizing gravity stormwater pipelines. The same simplified technique of using only peak flow values, however, cannot be applied to the design of stormwater pumping stations.

Table 2.1 Recommended wastewater flows from establishments served by a community wastewater system (Standards for Sewage, 1983).[a]

Type of establishment	Wastewater flow, gpd/person[b]
Residential	
Hotels and motels without private baths	40
Hotels and motels with private baths	50
Luxury residences and estates	125
Multiple-family dwellings and apartments	60
Rooming houses	50
Single-family houses	75
Commercial	
Airline catering per meal served	3
Airports (per passenger, not including food)	5
Airports (per employee)	10
Beauty shops (per operator chair)	200
Bus service areas, not including food	5
Country clubs, not including food	30
Drive-in theaters (per space per day, not including food)	10
Factories and plants (exclusive of industrial wastes)	35
Laundries, self-service (gallons/washer)	400
Mobilehome parks, independent (per space)	250
Movie theaters (per auditorium seat, not including food)	5
Offices (per employee)	10
Restaurants (per patron, toilet and kitchen wastes)	10
(Additional for bars and cocktail lounges)	2
Restaurants (single-service utensils/person, kitchen and toilet wastes)	8.5
Restaurants (single-service utensils/patron, kitchen waste only)	3
Stores (per public toilet)	400
Warehouses	35
Work or construction camps (semipermanent) with flush toilets	50
Work or construction camps (semipermanent) without flush toilets	35
Institutional	
Churches (per seat)	3
Churches (per meal served, kitchen waste)	3
Churches (per meal served, with paper service)	1.5
Hospitals (per bed space, with laundry)	300
Hospitals (per bed space, without laundry)	220
Institutional food service (per meal)	20
Institutions other than hospitals (per bed space)	125
Schools, boarding	100
Schools, day (without cafeterias, gyms, or showers)	15
Schools, day (with cafeterias, but no gym or showers)	20
Schools, day (with cafeterias, gyms, and showers)	25

Table 2.1 Recommended wastewater flows from establishments served by a community wastewater system (continued).[a]

Type of establishment	Wastewater flow, gpd/person[b]
Recreational and seasonal	
Camps, day (no meals served)	10
Camps, hunting and summer residential (night and day) with limited plumbing, including water-carried toilet wastes	50
Campgrounds, with individual sewer and water hookup (per space)	100
Campgrounds, with water hookup only and/or central comfort station that includes water-carried toilet wastes (per space)	50
Fairgrounds and parks, picnic with bathhouses, showers, and flush toilets	15
Fairgrounds and parks, picnic (toilet wastes only)	5
Swimming pools and bathhouses	10

[a] Actual flow data for any establishment indicating peak daily flows over a 1-year period may be accepted for use in sizing the onlot disposal system.

[b] $\text{gpd} \times 3.785 \times 10^{-3} = \text{m}^3/\text{d}$.

The design of stormwater pumping stations requires not only ascertaining the peak flow, but also defining the variation of flow with time to determine the design inflow to the pumping station. This curve of flow versus time is known as the inflow hydrograph. Methods for its development will be discussed later in this chapter. A distinction needs to be made that the stormwater hydrograph may not necessarily be the same as the pumping station inflow hydrograph. This is because the point of measurement for the stormwater event may be different from the location of the pump station. As described above for sanitary flows, an engineering design report also should be prepared for stormwater pumping stations. The report should address the system hydrology and the elements (such as existing and future watershed development, soil and cover type, and meteorological data) that will affect stormwater generation, methodology and development of the influent hydrograph, and selection of appropriate design flows.

INFLOW HYDROGRAPH. The inflow hydrograph must be defined to size properly and economically the pumps and pumping station wet well. For stormwater pumping stations, it is advisable to size pumps to accommodate the hourly peak inflow or provide variable pumping capacity to match the inflow hydrograph. A limited storage capacity must be provided in the pumping station wet well, but additional storage may serve to equalize the incoming flow and will reduce the required pumping capacity. The designer's objective is to establish a cost-effective balance between wet well storage volume and pumping capacity, and the possibility that the wet well will be partially full with limited equalization capacity should be considered as a worst case scenario.

Future development of the watershed should be considered and incorporated to derivation of the inflow hydrograph. If the watershed already is fully developed, and neither the impervious area nor the collection system will change significantly in the future, the inflow hydrograph for the pumping station may be based on existing conditions. However, if the watershed characteristics or collection system will be modified by future development, it is judicious to plan and design for both present and anticipated future conditions over the life of the pumping station. One common and accepted way of providing increased future capacity is to leave space for additional pumps in the station. When this is done, the capacity of the station inflow and discharge piping system, wet well hydraulics, and power and control system also should be designed for future requirements. However, care must be exercised so as not to oversize the pump suction pipes to the point of causing a gross mismatch between them and the pump suction. Extreme mismatches between the diameters could lead to inadequate hydraulic operation of the smaller, initial-year pump. Details covering pump suction piping are discussed in Chapter 5.

Methods of Developing Inflow Hydrographs. There are several generally accepted methods for calculating the inflow hydrograph, including those developed by the U.S. Bureau of Reclamation (USBR), the U.S. Soil Conservation Service (USSCS), and the U.S. Geological Survey (USGS). One method, still in wide use, is the rational method. Although a number of state and local agencies now prefer or require newer computerized methods of analysis, the rational method retains popularity for analysis of smaller, less complex systems because of its simplicity.

RATIONAL METHOD. The rational method of hydrologic analysis consists of an empirical formula that attempts to relate rainfall and runoff through use of a single coefficient. The coefficient must account for all losses in the watershed, including infiltration, evapotranspiration, and surface depression storage. The rational method is discussed extensively in *Design and Construction of Sanitary and Storm Sewers* (1969).

Numerous efforts have tried to improve the accuracy of estimates used for the equation's runoff coefficient, to the point that the use of the rational formula itself often is secondary. Moreover, because the rational method provides only an estimate of the peak runoff rate, additional methodology is applied to derive the runoff hydrograph.

The basic equation is

$$Q = C\,i\,A \tag{1}$$

Where

Q	=	the peak runoff rate, L/s (cu ft/sec);
C	=	the runoff coefficient reflecting characteristics of the drainage area and the collection system;
i	=	the average rainfall intensity, cm/h (in./hr); and
A	=	the drainage area, ha (ac).

The peak rate of runoff at any point is a direct function of the average rainfall intensity during the time of concentration to that same point. The frequency of the peak discharge is the same as the frequency of the average rainfall intensity. Finally, the time of concentration is the time required for the runoff to become established and flow from the most remote part of the drainage area to reach the point under design.

The last assumption applies to the part of the watershed most remote in time, not necessarily in distance. When using the rational method, average rainfall intensities have no time-sequence relationship to the actual rainfall pattern during the storm, and the intensity duration curve is not a time-sequence curve of actual precipitation. In addition, the rational method does not incorporate effects of preceding storm events and the way in which they may have altered the ground condition (such as saturation). Failure to recognize these points has led to improper usage of the rational method in attempts to evaluate the effects of antecedent precipitation and infiltration.

Use of the rational method should be limited to small watersheds [fewer than approximately 13 km^2 (5 sq miles)] (Linsley *et al.*, 1958). For larger watersheds, more reliable and accurate hydrograph development methods (such as those developed by USBR, USSCS, and USGS) are more appropriate. In these cases, a simple linear relationship with a single runoff coefficient cannot produce an adequate prediction of runoff, which involves storage, subsurface flow, stage discharge and storage, and phenomena previously discussed.

Determination of the variables used in the rational method (equation 1) are discussed in *Design and Construction of Sanitary and Storm Sewers* (1969). These variables, which also are used in other methods, include area, rainfall, and runoff.

The area tributary to any point under consideration in a storm collection system must be determined. This information should include the component parts of the drainage area, land use, soil characteristics, and general slopes in the topography.

Rainfall information includes intensity (i), duration, and frequency determination, all of which are interrelated. The average frequency of rainfall occurrence used for design determines the degree of protection afforded by a given storm sewer system. This protection should be consistent with the amount of damage prevented, thus involves an economic analysis of cost (prevention) and benefit (protection). In practice, detailed cost-benefit studies typically are not conducted for small, urban, storm drainage projects. Judgment supported by records of performance in other similar areas is typically the basis of selecting the design frequency.

The ranges of frequencies used are as follows:

- For storm sewers in residential areas, 2 to 15 years, with 5-year frequencies most common.
- For storm sewers in commercial and high-value districts, 10 to 50 years, depending on economic justification.
- For flood protection works, 50 years or more with 100 years common for many public works projects.

Note that in the latter two situations, a more detailed economic analysis frequently is justified because of the greater cost of providing additional protection.

Considerations that may lead to the use of other design storm frequencies include

- Potential for future relief from drainage problems in parts of the system (for example, detention or flow equalization);
- Combined sewer systems that warrant using higher design rainfall frequencies;
- Special structures, such as pumping stations, that justify higher design storm frequencies;
- Lack of funding to provide the level of protection desired; and
- Requirements of regulatory agencies (for example, flood insurance program requirements and soil conservation).

Typically, the cost of providing additional protection is low compared to potential damage from flooding. For stormwater pumping stations, at least a brief economic analysis of a more conservative rainfall frequency design usually is justified. Figure 2.1 illustrates a typical economic analysis obtained by plotting both the costs because of damage from flooding or inadequate pumping capacity versus costs of providing increased pumping capacity. The sum of these two costs generally will yield a minimum point that indicates the minimum cost balance.

Basic rainfall data are derived from gauge measurements of rainfall. Length of record, gauge accuracy and placement, and density significantly affect the reliability of rainfall data.

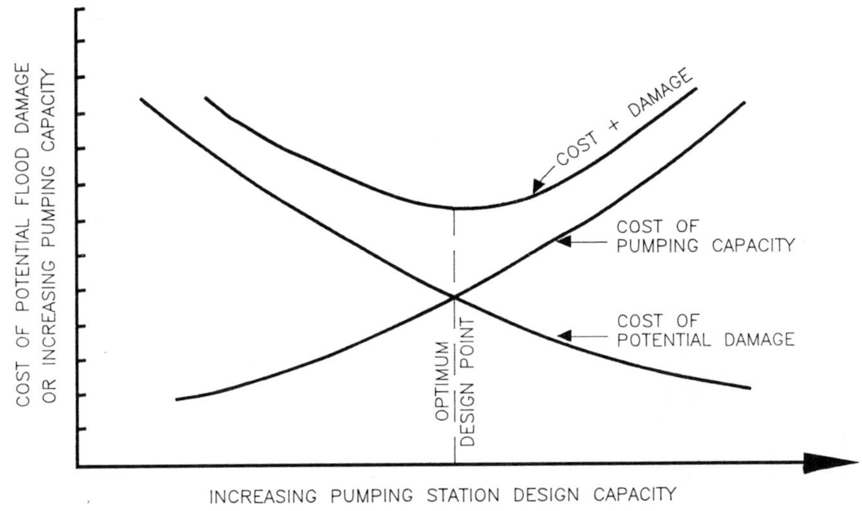

Figure 2.1 Typical cost optimization curves based on potential flood damage compared to increased pumping capacity.

It is important to note the relationship between total accumulated rainfall versus the length of the storm event and the return frequency (Generalized Estimates, 1960; Frederick *et al.*, 1977; Rainfall Frequency, 1961; and Rainfall Intensities, 1956). Longer duration storm events generally will have a greater total accumulation because the rainfall is accumulated over a longer period but may have a lower peak intensity (instantaneous intensity). Likewise, the longer the return period the greater the total accumulation. Rainfall intensity-duration curves for the selected locality and frequency may be prepared from these and similar charts. Figure 2.2 illustrates a curve of this type.

The relationship of rainfall intensity duration, storm pattern, time of concentration, and average intensity is discussed in *Design and Construction of Sanitary and Storm Sewers* (1969). An estimate of the time of concentration to the point under consideration is made so that the average rainfall rate for the design period can be determined. For urban storm sewers, the time of concentration consists of the inlet time plus the time of flow in the sewer from the most remote inlet to the point under consideration.

Decreasing point rainfall intensity values for large areas may result in lower average intensities. Figure 2.3 (Rainfall Frequency, 1961) is based on data gathered from 20 dense networks of gauges in various regions of the U.S. east of the 90th meridian and gives factors for converting the average point rainfall intensity to the average intensity over the area of interest.

Care must be exercised when working with irregularly shaped drainage areas. Occasionally, a part of the area having a shorter time of concentration—and thereby subject to a higher-intensity rainfall—may cause a greater runoff

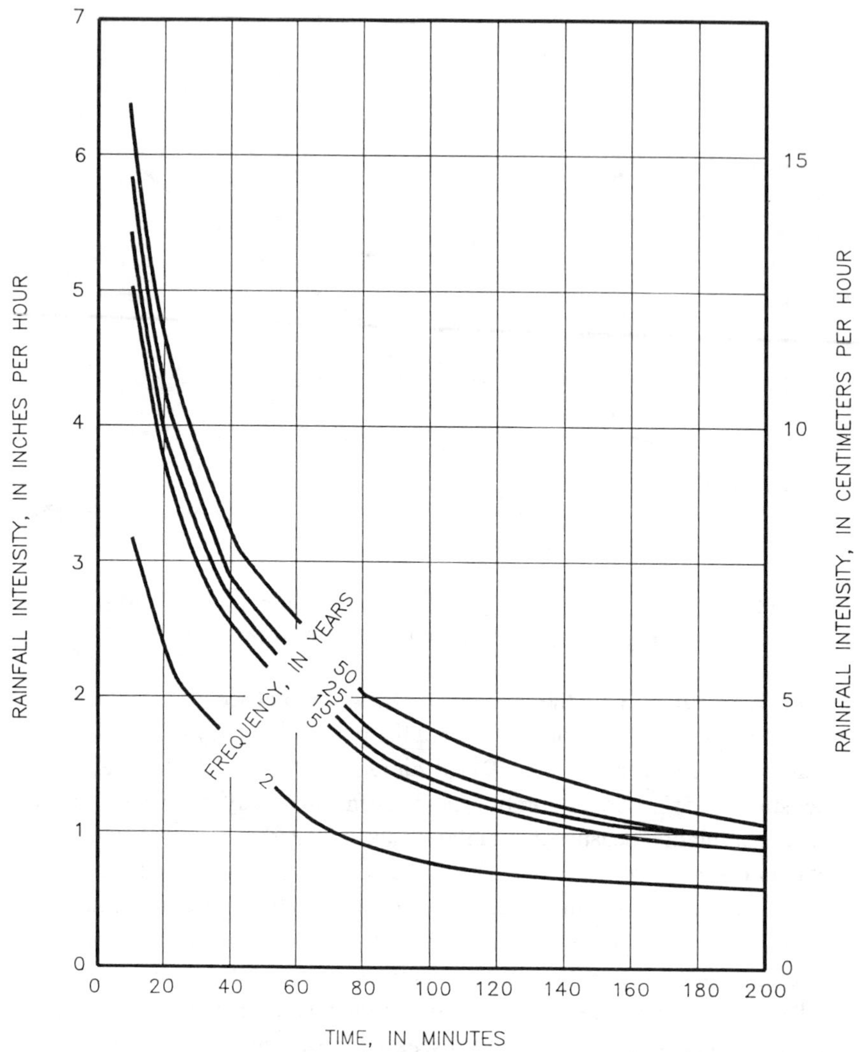

Figure 2.2 Intensity-duration rainfall curves, Boston, Massachusetts.

rate at a design point than that contributed by the entire area with its longer time of concentration and lower rainfall intensity.

The runoff coefficient (*C*) is the variable of the rational method that is hardest to determine. Its use in the formula implies a fixed ratio between rainfall intensity and resultant runoff for any given drainage area. In reality, the coefficient also should account for abstractions or losses between the rainfall and runoff volumes, which may vary for a given drainage area and that are influenced by differing hydrological, climatological, and seasonal conditions. These losses include interception by vegetation, infiltration to permeable soils, retention in surface depressions and impoundments, and transpiration.

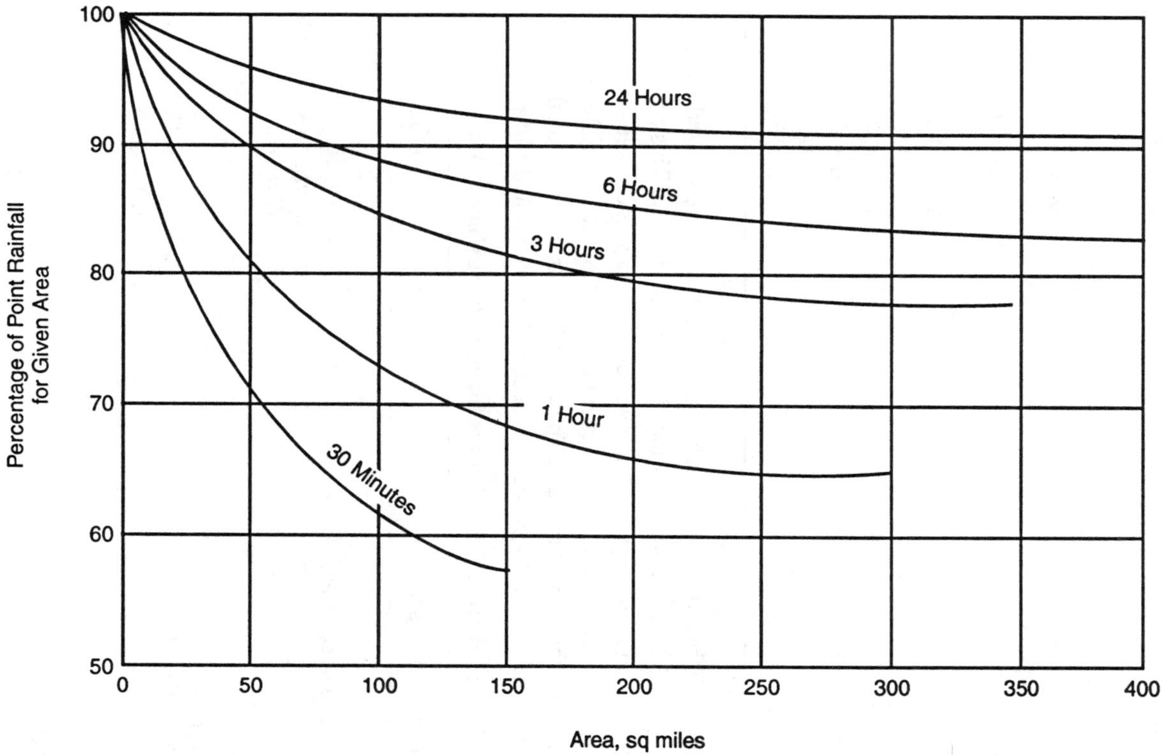

Figure 2.3 **Converting average point rainfall to the average over the area of interest (area–depth curves) (sq miles × 2.59 = km^2) (Rainfall Frequency, 1961).**

The use of average coefficients for various surface types, which are assumed not to vary through the duration of the storm, is common. The range of coefficients, clarified with respect to the general character of the tributary area reported in use, is given in Table 2.2.

Table 2.2 Range of runoff coefficients for various tributary areas.

Description of area	Runoff coefficients
Business	
Downtown	0.70–0.95
Neighborhood	0.50–0.70
Residential	
Single-family	0.30–0.50
Multiunits, detached	0.40–0.60
Multiunits, attached	0.60–0.75
Apartment	0.50–0.70
Industrial	
Light	0.50–0.80
Heavy	0.60–0.90
Parks, cemeteries	0.10–0.25
Playgrounds	0.20–0.35
Railroad yard	0.20–0.35
Unimproved	0.10–0.30

Often, it is desirable to develop a composite runoff coefficient based on the percentage of different kinds of surface in the drainage area. This procedure frequently is applied to typical "sample" blocks as a guide to selecting reasonable values of the coefficient for an entire area. Coefficients reflecting surfaces in use are shown in Table 2.3. These coefficients are based on the assumption that the design storm does not occur when the ground surface is

Table 2.3 Range of runoff coefficients for various surface types.

Character of surface	Runoff coefficients
Pavements	
Asphaltic and concrete	0.70–0.95
Brick	0.70–0.85
Roofs	0.75–0.95
Lawns, sandy soil	
Flat, 2%	0.05–0.10
Average, 2 to 7%	0.10–0.15
Steep, 7%	0.15–0.20
Lawns, heavy soil	
Flat, 2%	0.13–0.17
Average, 2 to 7%	0.18–0.22
Steep, 7%	0.25–0.35

frozen and impermeable, yet this condition existed during some flood disasters. Therefore, this assumption is a weakness of the rational method.

To apply the rational method to the design of stormwater pumping stations, it is necessary to develop an inflow hydrograph (Williams, 1950). By itself, the rational method produces only a peak flow that is assumed to be constant once the entire watershed is contributing to the flow (that is, the time of rainfall is greater than the time of concentration). Several simplified methods have been proposed to develop an inflow hydrograph from the peak flow produced by the rational method. One of the most common methods is the development of a triangular hydrograph where the peak flow is the output from the rational method, and the base of the hydrograph is made up of the time-to-peak (*Tp*) and the time-of-recession (*Tr*) (Figure 2.4) (Design of Small Dams, 1965).

For most ungauged small watersheds, the Soil Conservation Service has shown that the time for a storm event hydrograph to cease after peaking (*Tr*) is related approximately to the time-to-peak (*Tp*) as follows:

$$Tr = 1.67Tp \qquad (2)$$

Such simplified approaches to storm event analysis should be applied only to small watersheds of fewer than 13 km^2 (5 sq miles). Even then, any available information affecting runoff characteristics also should be considered. An attempt should be made to balance the excess rainfall with the volume of runoff water represented by the area under the triangular hydrograph.

SOIL CONSERVATION SERVICE METHOD. The Soil Conservation Service (SCS) developed methods to be applied in estimating the runoff hydrograph for small urban watersheds (Urban Hydrology, 1975; and SCS, 1975). Following are parameters involved in the SCS method.

Rainfall. The SCS uses five standard rainfall distributions: types I, IA, II, III, and IV. Type II generally applies to all areas of the U.S. except for parts of the Pacific Coast states. For rainfall distribution in the Pacific Coast states, the map in Appendix D of "Urban Hydrology for Small Watersheds" (1975) should be used.

Runoff Volume Parameters. Because urban areas seldom are covered completely by impervious structures, soil properties are important in estimating the total volume of direct runoff. Infiltration and percolation rates of soils indicate their potential to absorb rainfall and thereby reduce the amount of direct runoff.

The type of cover and its hydrologic condition affect runoff volume through its influence on the infiltration rate of the soil. Fallow land yields

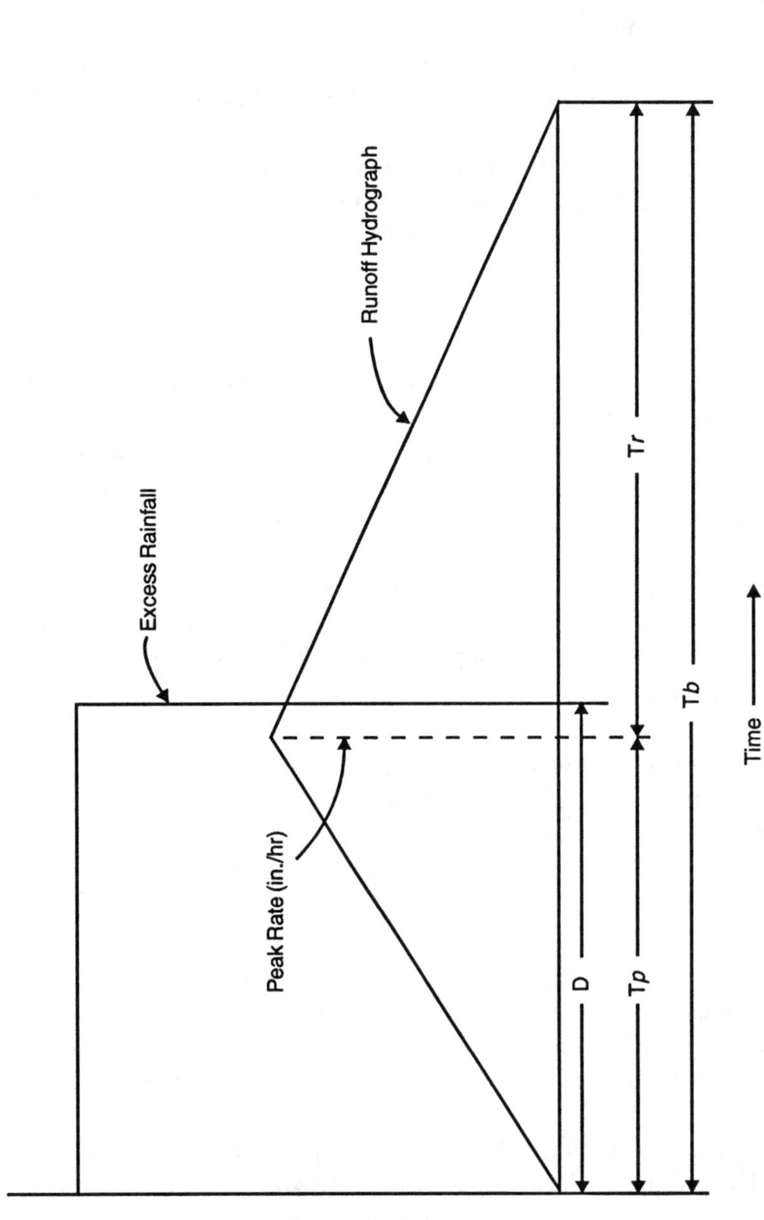

Figure 2.4 **Triangular hydrograph analysis: *Tp* = time in hours from start of rise to peak rate; *Tr* = time in hours from peak rate to end of triangle; *D* = rainfall excess period, hours; and *Tb* = time base of hydrograph (cu ft/sec × 28.32 = L/s, in. × 25.4 = mm, and in./hr × 25.40 = mm/h) (Design of Small Dams, 1965).**

more runoff than forested land for a given soil type. Covering areas with impervious material reduces surface storage and infiltration and increases the volume of runoff.

Some rainfall is retained on the surface and by vegetation before runoff begins. Interception is rainfall that is caught (for example, by foliage, twigs, branches, and leaves). This rainfall is lost to evaporation and never reaches the ground surface. Increasing the vegetal cover increases interception.

Surface depression storage begins when precipitation exceeds infiltration. Overland flow starts when surface depressions are full. The water captured in depression storage is not available as direct runoff.

Initial abstraction, which is the sum of interception, depression storage, and infiltration before runoff begins, occurs on all types of cover, from pasture in good condition to concrete pavement. The amount of initial abstraction is less on concrete pavement than on pasture.

Time Parameters. The slopes of storm sewers, street gutters, roads, overland flow areas, and stream channels are significant in determining travel times through urban watersheds.

The flow length is the distance the runoff must travel from the most distant portion of the watershed to the pumping station. This is not necessarily a straight line because flow must follow the topography.

The degree of surface roughness influences the time of runoff. Rough surfaces include woodland, grassland, and natural channels. Smooth surfaces include paved surfaces and concrete-lined channels.

Following is a brief outline of the methodology used in the SCS urban runoff method (SCS, 1975), which is recommended for drainage areas up to 52 km^2 (20 sq miles). Peak discharge and total direct runoff techniques may be used for areas up to 800 ha (2 000 ac) (Urban Hydrology, 1975).

Determination of Runoff Volume. Runoff volume is derived from information on the rainfall amount and watershed characteristics. A combination of a hydrologic soil group, land use, and treatment class is used to determine the hydrologic soil-cover complex. The effect of the hydrologic soil-cover complex on the volume of rainfall runoff is represented by a runoff curve number (CN). If the rainfall amount and CN are known, the volume of runoff from the watershed can be estimated.

Time of Concentration, Travel Time, and Lag. As with the rational method, "time of concentration" is the time it takes for runoff to travel from the hydraulically most distant part of the watershed to the point of reference and is computed by determining the water travel time through the watershed. In hydrograph analysis, it is the time from the end of excessive rainfall to the point of inflection on the falling limb of the hydrograph. Lag can be considered a weighted time of concentration and is related to the physical properties of a watershed, such as area, length, and slope. In simple hydrograph analysis, lag is the time from the center of mass of an excessive rainfall to the peak rate of runoff. The time of concentration determines the shape of the runoff hydrograph. Thus, changes in the time of concentration cause changes in the resulting hydrograph. The extent of urbanization and stream modification affects the travel time of water through the watershed. Travel time, in turn, changes the time of concentration.

Travel time and time of concentration can be determined using available tables and equations for overland flow, channel flow, and storm sewer flow. Lag time typically is calculated as a proportion (typically 0.7) of the time of concentration.

Peak Discharge. After the time of concentration and travel time, the drainage area, the runoff curve number, the 24-hour rainfall and its distribution, and the runoff in inches are known, the peak discharge can be obtained from SCS tables (Urban Hydrology, 1975).

COMPUTERIZED RUNOFF MODELS. Over the past decade, computer models have been developed to analyze urban runoff quantity and quality. Most of these models are available in versions compatible with personal computers and have been adapted for user friendly interface, easy data input, and graphical presentation of results. Private software developers have integrated the more popular runoff models with computer graphics and plotting software to enhance the interpretation and presentation of results.

Computer models are used to study quantity and quality problems associated with urban stormwater runoff in which three broad system objectives can be identified: planning, design, and operation. Typically, each objective will require a model with somewhat different, yet overlapping, characteristics. It should be noted that all models are based on a variety of assumptions, and the user must understand their nature and proper use to avoid erroneous model results.

Planning Models. Planning models can provide an overall assessment of the urban runoff problem and an estimate of the effectiveness and cost of abatement procedures. They are often used for an initial analysis of the rainfall–runoff relationship and may illustrate tradeoffs among various control options (for example, treatment versus storage). These models have relatively large time steps (hours) and long simulation times (months or years). Data requirements typically are minimal and their mathematical complexity is low. A good example is the storage, treatment, overflow, and runoff model (STORM) developed by the Corps of Engineers Hydrologic Engineering Center (HEC) (Huber *et al.*, 1975, and Urban Storm Water, 1974).

Design Models. Design models offer the user a more detailed simulation of a single storm event. Most of the current design models provide a complete description of flow and pollutant routing from the area of rainfall through the entire collection system and often to the receiving waters as well. Design models can be used to predict flows and concentrations anywhere in the rainfall and collection system and can illustrate more clearly how abatement procedures or design options affect them. These models are a rapid and cost-effective means of determining least-cost abatement procedures for both quantity and quality problems in urban areas.

Design models are used to simulate a single storm event and have short time steps (minutes) and short simulation times (hours). Data requirements may be moderate to extensive, depending on the particular model employed, and mathematical analysis is more complex than at the planning level.

The U.S. EPA Storm Water Management Model (SWMM) is an example of a design-level model developed to simulate urban stormwater quantity and quality processes and to assess the effectiveness of alternative abatement and mitigation procedures (Huber *et al.*, 1977).

Operational Models. Operational models are used either to produce actual control decisions or to recommend a control strategy to operating personnel during a real storm event. For real-time operational models, rainfall data are telemetered from remote stations and entered directly to the model, which then is used to predict collection system responses a short time into the future. Various control options then can be employed (for example, in-system storage, diversions, or regulator settings).

Operational models are developed from existing sophisticated design models that have been adapted and calibrated for a particular system. Examples are those designed for Minneapolis–St. Paul (Bowers *et al.*, 1968), Seattle (Leiser, 1974), and the Washington Suburban Sanitary Commission (Sewer System, 1983).

COMBINED FLOW SYSTEMS

Combined sewer systems are those collection, conveyance, and pumping systems designed to carry both sanitary flow and stormwater flow in a single system. Although few combined sewer systems now are being constructed, many cities have extensive existing combined sewer systems dating from a number of years ago. For reasons including cost, politics, and feasibility, it is unlikely these systems will be modified to accommodate segregated flows in the foreseeable future. For that reason and because the engineer often must expand or modify a combined sewer system, a brief description of the methods used to determine combined flows is presented.

The design of combined sewers is the same as for storm sewers, though the design of pumping stations for combined flows must consider wastewater flows and volumes. Long detention times in a wet well cause septic conditions and odor problems. Thus, to determine the capacity for a combined flow pumping station, both sanitary wastewater and stormwater flows must be considered. For low flow purposes (for example, dry weather conditions) wastewater rates are considered alone. Pumping capacity and wet well volume should be designed to provide adequate but short detention times for wastewater. For higher, wet weather flows, the sanitary wastewater flow can be added to the stormwater flow (or neglected altogether if insignificant), and additional pumps in the same station can be sized to handle these large stormwater flows.

Bypass and overflow devices on the collection system should be considered when determining system flows and designing pumping station capacities. Design examples illustrating the calculation of stormwater flows, wastewater flows, and combined flows are presented in *Wastewater Engineering* (Metcalf and Eddy, 1991).

INFILTRATION AND INFLOW

When calculating design flows for pumping stations, consideration must be given to infiltration and inflow. Infiltration is ground water that enters a sewer through defective and cracked pipes, open or loose pipe joint connections, and leaking manhole walls. Inflow is water entering a sewer system and service connections from roof drain leaders, cellar drains, cooling water discharges, manhole covers, cross connections, and surface runoff (Design of Municipal, 1992, and Existing Sewer, 1983).

Since the implementation of the Federal Water Pollution Control Act Amendments of 1972, which required grant recipients to demonstrate their sewer systems were not subject to excessive I/I, emphasis has continued to be placed on assessing and limiting I/I to sanitary sewer collection systems. While a cost-benefit analysis ultimately will determine whether the owner of a system should rehabilitate a sewer with excessive infiltration/inflow, the designer of a pumping station needs to be able to assess the impact these additional flows will have on the design of the pumping facilities and ultimate treatment facilities.

In the past, and commonly in older collection systems, I/I have contributed large additional flows to storm and sanitary sewers, especially during wet weather periods when ground adjacent to the sewer pipe is saturated. Minimum rates of infiltration, based on complete submergence at varying depths, are presented in Table 2.4 (Metcalf and Eddy, 1991). The advantages of new pipe joint materials, such as rubber gasket and new-style polyvinyl chloride, are evident.

Table 2.4 Submergence test results showing infiltration ratio for 6-in. diameter clay pipe and concrete pipe.[a]

6-in. diameter clay pipe			6-in. diameter concrete pipe		
Joint material	**Head above flow line, in.**	**Average infiltration rate, gpd/in. dia/mile**	**Joint material**	**Head above flow line, in.**	**Average infiltration rate, gpd/in. dia/mile**
Jute only	3	8 270	Jute only	3	6 710
	9	71 050		9	52 800
	15	155 250		15	118 000
	21	258 000		21	205 500
	27	356 000		27	278 000
Cement	3	3 360	Cement	3	680
	9	15 000		9	4 950
	15	28 700		15	10 450
	21	41 200		21	16 500
	27	53 200		27	22 000
Hot pour	3	1 330	Cold mastic	3	990
	9	1 660		9	1 450
	15	3 410		15	3 210
	21	4 720		21	5 130
	27	5 520		27	7 810
Old-style polyvinyl chloride	3	0	Hot pour	3	0
	9	645		9	107
	15	1 450		15	235
	21	1 850		21	419
	27	2 400		27	513
New-style polyvinyl chloride	0–27	Negligible	Rubber gasket	0–27	Negligible

[a] in. × 25.4 = mm; gpd/in. dia/mile × 92.6 = (mL/m·d)/m.

Infiltration and inflow to sanitary sewers most often is estimated in terms of the amount of water entering a sewer based on the diameter and length of pipe, (mL/m·d)/m (gpd/in. dia/mile). Infiltration may range from 9 260 to 926 000 (mL/m·d)/m (100 to more than 10 000 gpd/in. dia/mile) (Gravity, 1982). The actual amount of I/I will depend on the material and integrity of the sewer system, the number and type of building connections, the elevation of the ground water compared to that of the sewers, the integrity of manholes, and the degree of manhole submergence during a rainfall event.

Approximate design allowances for I/I can be determined from Figure 2.5 (Metcalf and Eddy, 1991). Curves A and B are used for older sewers, and selection is based on the relative condition of the system. Curve C is used for

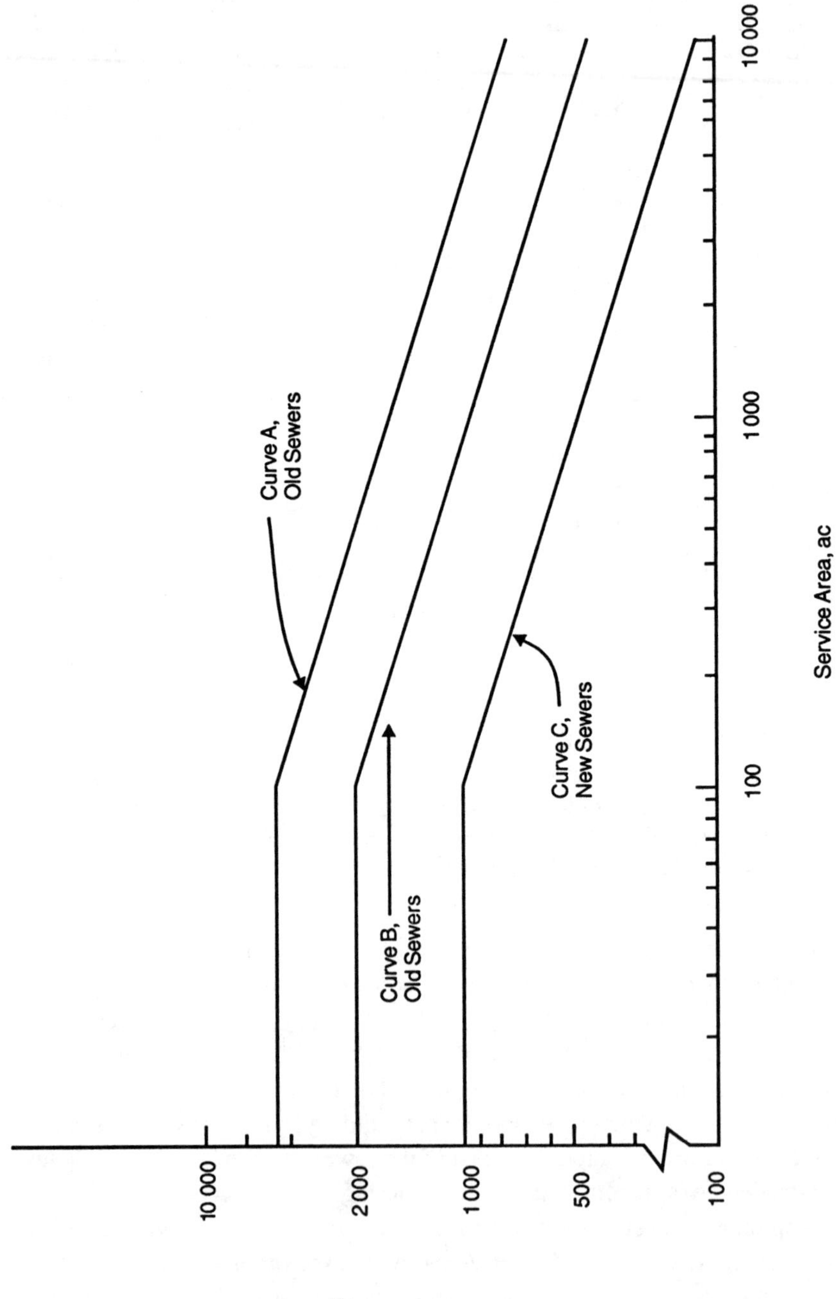

Figure 2.5 Infiltration rate curves for old and new sewers (gpd/ac × 9.353 = L/ha·d).

new sewers where the ground water is never above the pipe invert and the inflow of extraneous water is low. Detailed pumping station design should rely on a more complete sewer system I/I evaluation that establishes accurate I/I flow rates through actual field measurement.

FLOW REGULATION

To a limited degree, the pumping station wet well volume may be used to equalize inflow to the pumps, and thus reduce the size or number of pumps required. Although theoretically this storage can be used for both stormwater and wastewater pumping stations, long storage periods in wastewater pumping stations could be impractical because of the problems of solids settlement and odor production.

To size the wet well for a stormwater pumping station, the inflow hydrograph must be routed through the wet well. The general equation for routing a flow through any storage facility is

$$\text{Rate of inflow} - \text{rate of outflow} = \text{rate of change of storage} \qquad (3)$$

The method of reservoir routing is covered in most hydrologic texts (Chow, 1964; Domestic Wastewater, 1991; and Linsley *et al.*, 1958). In a pumping station, the rate of inflow is derived from the inflow hydrograph, and the rate of outflow is the pumping rate. The rate of change of storage is a direct function of the shape of the entrance to the pump station and wet well. The piping system leading to the pumping station may also contribute to storage, depending on its invert elevation relative to that of the wet well. When available, this latter storage may be significant.

The total pump capacity must be balanced with the storage volume provided to obtain the most economical balance between the two. The method of mass curves also can be applied to the graphical solution of the routing equation (Linsley and Franzini, 1972).

QUALITY OF INFLUENT

The quality characteristics of the influent are as important as the quantity and type of influent to the pump station. These quality characteristics will affect the overall design of the proposed facilities, influence the selection of the equipment, and set certain required operational procedures once constructed. The following types of characteristics should either be determined from existing information or a reasonable judgment made of their likely presence in the influent (Sanks *et al.*, 1991):

- Pumps should be sized to accommodate the largest solids, if at all possible. In the case of storm or combined flows, it may be necessary to add mechanical screening or grinding to the facilities design. It is common practice to protect wastewater pumps from solids and rags by installing comminutors or grinders in influent channels before the wet well. These additional items will increase facility cost and have a large impact on operation and maintenance costs.
- Unusual grit quantities could determine special materials for pump construction or the need for upstream grit removal.
- Corrosive constituents, particularly when dealing with industrial waste flows.
- The presence of toxic, explosive, or hazardous materials may establish requirements for monitoring and special venting.
- Possibility of septic wastewater and large quantities of hydrogen sulfide or other odor-causing substances. These characteristics may require the use of special coatings to reduce corrosion, venting, and odor control for operator safety and of variable-speed pumping equipment to reduce residence wet well time.

REFERENCES

Bowers, C.E., *et al.* (1968) The Real Time Computation of Runoff and Storm Flow in the Minneapolis–St. Paul Interceptor Sewers. Memo No. M-118, St. Anthony Falls Hydraul. Lab., Univ. Minn., Minneapolis.

Chow, V.T. (1964) *Handbook of Applied Hydrology.* McGraw–Hill, Inc., New York, N.Y.

Chow, V.T., *et al.* (1988) *Applied Hydrology.* McGraw–Hill, Inc., New York, N.Y.

Design and Construction of Sanitary and Storm Sewers (1969). Manual of Practice No. 9, Water Pollut. Control Fed., Washington, D.C.; Manuals and Rep. of Eng. Practice No. 37, Am. Soc. Civ. Eng., New York, N.Y.

Design of Municipal Wastewater Treatment Plants (1992). Manual of Practice No. 8, Water Pollut. Control Fed., Washington, D.C.; ASCE Manuals and Rep. on Eng. Practice No. 76, Am. Soc. Civ. Eng., New York, N.Y.

Design of Small Dams (1965). U.S. Dep. of the Interior, Bureau of Reclamation, Third Printing, U.S. Gov. Printing Office, Washington, D.C.

Domestic Wastewater Facilities Manual (1991). Pa. Dep. Environ. Resour., Bureau of Water Qual. Manage.

Existing Sewer System Evaluation and Rehabilitation (1983). Manual of Practice No. FD-6, Water Pollut. Control Fed., Washington, D.C.; Manuals and Rep. on Eng. Practice No. 62, Am. Soc. Civ. Eng., New York, N.Y.

Frederick, R.H., *et al.* (1977) Five-to-60-Minute Precipitation Frequency for the Eastern and Central United States. NOAA Tech. Memo. NWS HYDRO-35, Office of Hydrol., Natl. Weather Serv., Natl. Oceanic and Atm. Adm., Silver Spring, Md.

Generalized Estimates of Probable Maximum Precipitation for the United States West of the 105th Meridian for Areas to 400 Sq Miles and Durations to 24 Hr. (1960). Weather Bureau, Tech. Paper No. 38, U.S. Gov. Printing Office, Washington, D.C.

Gravity Sanitary Sewer Design and Construction (1982). Manual of Practice No. FD-5, Water Pollut. Control Fed., Washington, D.C.; Manuals and Rep. on Eng. Practice No. 60, Am. Soc. Civ. Eng., New York, N.Y.

Huber, W.C., *et al.* (1975) Storm Water Management Model—User's Manual Version II. Univ. Fla., Gainesville.

Huber, W.C., *et al.* (1977) Interim Documentation, November 1977 Release of EPA SWMM. U.S. EPA Draft Rep., Proj. No. R-802411, Natl. Environ. Resour. Cent., Cincinnati, Ohio.

Leiser, C.P. (1974) Computer Management of Combined Sewer System. EPA-670/2-74-022, U.S. EPA, Washington, D.C.

Linsley, R.K., Jr., *et al.* (1958) *Hydrology for Engineers.* McGraw–Hill, Inc., New York, N.Y.

Linsley, R.K., Jr., and Franzini, J.B. (1972) *Water Resources Engineering.* McGraw–Hill, Inc., New York, N.Y.

McJunkin, F.E. (1964) Population Forecasting by Sanitary Engineers. *J. Sanit. Eng. Div., Proc. Am. Soc. Civ. Eng.*, **90**, SA4.

McLean, J.E. (1952) More Accurate Population Estimate by Means of Logistic Curves. *Civ. Eng.*, **22.**

Metcalf and Eddy, Inc. (1991) *Wastewater Engineering.* 3rd Ed., McGraw–Hill, Inc., New York, N.Y.

"Population Projections by the Fertility Rate Method." A U.N. Publication, New York, N.Y.

Rainfall Frequency Atlas of the United States for Durations from 30-Min to 24-Hr and Return Periods from 1 to 100 Yr (1961). Weather Bureau, Tech. Paper No. 40, U.S. Gov. Printing Office, Washington, D.C.

Rainfall Intensities for Local Drainage Design in Western United States, for Durations of 20-Min to 24-Hr and 1 to 100-Yr Return Periods (1956). Weather Bureau, Tech. Paper No. 28, U.S. Gov. Printing Office, Washington, D.C.

Sanks, R.L., *et al.* (1991) *Pumping Station Design.* Butterworths, Stoneham, Mass.

SCS National Engineering Handbook, Section 4, Hydrology (1975). U.S. Dep. Agric., Soil Conserv. Serv., Washington, D.C.

Sewer System Computer Model: Design Report, Vols I, II, III, and IV (1983). Unpublished Eng. Rep., Wash. Suburban Sanit. Comm., Washington, D.C.

Standards for Sewage Disposal Facilities (1983). Pa. Dep. Environ. Resour. Rules and Regulations.
Urban Hydrology for Small Watersheds (1975). Tech. Release No. 55, U.S. Dep. Agric., Soil Conserv. Serv., Eng. Div., Washington, D.C.
Urban Storm Water Runoff: STORM (1974). Hydrol. Eng. Cent., Corps of Eng., Davis, Calif.
Williams, G.R. (1950) Hydrology. In *Engineering Hydraulics.* John Wiley and Sons, Inc., New York, N.Y., 309.

Chapter 3

Station Configuration and Design

This chapter discusses the configuration of different sizes and types of pumping stations and factors influencing the design of pumping stations, which include the needs of the end-user and satisfying environmental, social, and political constraints.

Station capacity, as described in Chapter 2, establishes the range of flows to be pumped, from which the number and size of pumps can be determined. Before detailed plans and specifications are prepared, a preliminary design layout should be presented to end-users, with scaled sketches showing number of pumps; pump type, selection, and arrangement; provisions for appurtenances; and site dimensions. Station configuration also can be affected by fluid characteristics; site constraints such as accessibility, maintenance location, and zoning ordinances; and appurtenances such as bar screens or comminutors, which may be needed to keep large objects from flowing down the sewer into the pumping station wet well. The pumping station and force main should be designed as a unit.

Because technical, environmental, and social and political constraints are unique to each facility, not all pumping stations are identical. This chapter provides guidance for the design of facilities to satisfy varying needs. Multiple design disciplines, including civil, structural, mechanical, and electrical engineering and instrumentation, are required for a successful design.

LOCATION CONSIDERATIONS

In determining the need for a wastewater or stormwater pumping station, the hydraulics of a sewer collection system must be evaluated. If gravity flows can be achieved economically, as determined by comparing the cost of constructing the pumping station and annual operating and maintenance expenses to the construction cost of gravity sewers, the station should be eliminated. Location and size of a pumping station should be determined after an analysis of technical needs, environmental constraints, local zoning ordinances and restrictions, economics, and public input during the planning stage. In developed areas, the "not-in-my-backyard" syndrome is an important factor in site selection.

The preliminary layout of a station depends on the design capacity, type of station, number, type, and the immediate and future capacity of pumps. Need and flexibility for future expansion also should be evaluated. If the station has a wide range of flows, the provision for a second future force main will affect station configuration but offers a wider range of pump combinations and selections. Other considerations in locating a pumping station include land ownership, accessibility, site and surrounding conditions such as flood potential and drainage, and availability of utilities (water, electricity, natural gas, and telephone). Also, subsoil and ground-water conditions and the necessary station depth of the specific substructure must be studied during site selection.

Site design should consider pathways of radio telemetry with central or relay stations (for alarm signals). If possible, stormwater pumping stations should be situated where water can be impounded with minimal potential of flood damage should inflow exceed station capacity. Impounding a portion of the stormwater inflow will control the rate of stormwater flow, consequently reducing the overall pumping capacity required for any given storm.

The selected station site should consider construction stage requirements, interceptor sewer and force main alignments, environmental impacts, vehicular access, site grading, future removal and replacement of equipment, noise constraints, and architectural and aesthetic needs. Malfunction of a wastewater pumping station can cause considerable public nuisance and property damage; therefore, reliability of power supply (including standby power) and control systems in addition to selection of appropriate pumping equipment is essential. The available types of drives and controls for pumping equipment and reliability of power supplies are discussed in Chapters 4, 6, and 7.

For small- to medium-capacity pumping stations, yard piping should be designed with bypass valving such that during total station failures, a transportable pumping unit such as engine-driven pumps can be brought in to provide emergency pumping. For larger pumping stations, use of transportable pumping units during emergencies may not be possible, so station reliability should be improved by installing standby pumps and generators. For rehabilitating existing pumping stations without standby power, the yard piping feature could be used. For new pumping stations, standby power is essential because of severe potential administrative fines against wastewater spillage.

Because of the location of gravity sewers, many pumping stations are located within flood plains, floodway fringes, and floodways, and thus are prone to flooding. To protect pumping station structures and electrical and mechanical equipment from flood damage, site the operating floor (where electrical and control panels typically are located) above the flood protection elevation (FPE) or design the pumping station against flooding. Though states or cities have different requirements, the FPE is considered the elevation 0.3 to 0.4 m (1 to 1.5 ft) above the maximum level flood plain. All entrances and unsealable openings in the station should be at elevations above the maximum flood level. Also, it is recommended that pumping stations be designed to remain fully operational and accessible during periodic high flood conditions. These criteria should be verified with local regulations and ordinances. For streams with no delineated maximum level flood plains, the design engineer should use appropriate computer models, such as the Hydrologic Engineering Center program developed by the U.S. Army Corps of Engineers, to determine the approximate maximum level flood plain elevation.

STATION TYPES

The type of pumping station selected should be dictated by such factors as design capacity, function, type of pumps needed, cost, and environmental and aesthetic considerations. The equipment components needed in a particular station to satisfy function and design requirements (such as screening, odor control, standby power, monorail, and hoist) may determine the station configuration and type. While design pumping capacity is the most significant variable in selecting the type of pumping station, size of site and site limitations may also affect the type of station chosen.

- Construction type
  - Prefabricated with package pumping equipment and controls;
  - Custom built or built-in-place; and
  - Packaged equipment for installation in manholes or built-in-place underground structures.

- Pump location
  - — Inside wet well;
  - — Inside dry well;
  - — Above ground; and
  - — Inclined shaft (screw pumps).
- Motor and drive location
  - — Above ground inside super structure;
  - — Below ground level inside dry well; and
  - — Below ground level inside wet well (integral with most submersible pumps).
- Motor and drive type
  - — Use of close-coupled, extended shafts or submersible motors could affect station configuration.
- Above ground equipment location
  - — Enclosed indoors and
  - — Exposed outdoors.

RANGES OF FLOW. Compared with most wastewater pumping stations, stormwater stations typically handle larger flows. To classify stations for handling either wastewater or stormwater, several ranges of flow can be used to distinguish several station types.

Small flow stations [up to 30 L/s (500 gpm)] can handle from a few houses to perhaps 500 or more residential buildings. Prefabricated package pumping stations usually should be considered for such applications, as should built-in-place stations with package pumping equipment. A single ejector or pump may be permitted, but duplex pumping units are recommended for system reliability, with the second pump providing 100% standby capacity in the event of lead-pipe failure.

Figures 3.1 and 3.2 illustrate general configurations of package pumping stations. Figure 3.1 shows duplex submersible pumping units in a single wet-well sump, ideal for handling wastewater or stormwater for a single building or a group of buildings where blockage by large objects is unlikely. Figure 3.2 shows dry-pit duplex pumping units inside a precast or prefabricated station, with the wet well being a separate but attached structure. Accessibility, safety, and future expansion needs should be considered for this type of design.

Comminutors have been used to reduce the size of solids in the flow prior to passing through the downstream pump. If larger solids in the flow stream necessitate a comminutor, then a combination grinder pump and bar screen should be considered.

For the small stations, non-clog-type pumps capable of passing 65- to 80-mm (2.5- to 3-in.) solids or grinder pumps are required. Screening devices to prevent clogging of stormwater pumps typically are warranted, and the duplex pumps should be designed to handle 100% of peak design flow, with the second pump serving as a full standby unit.

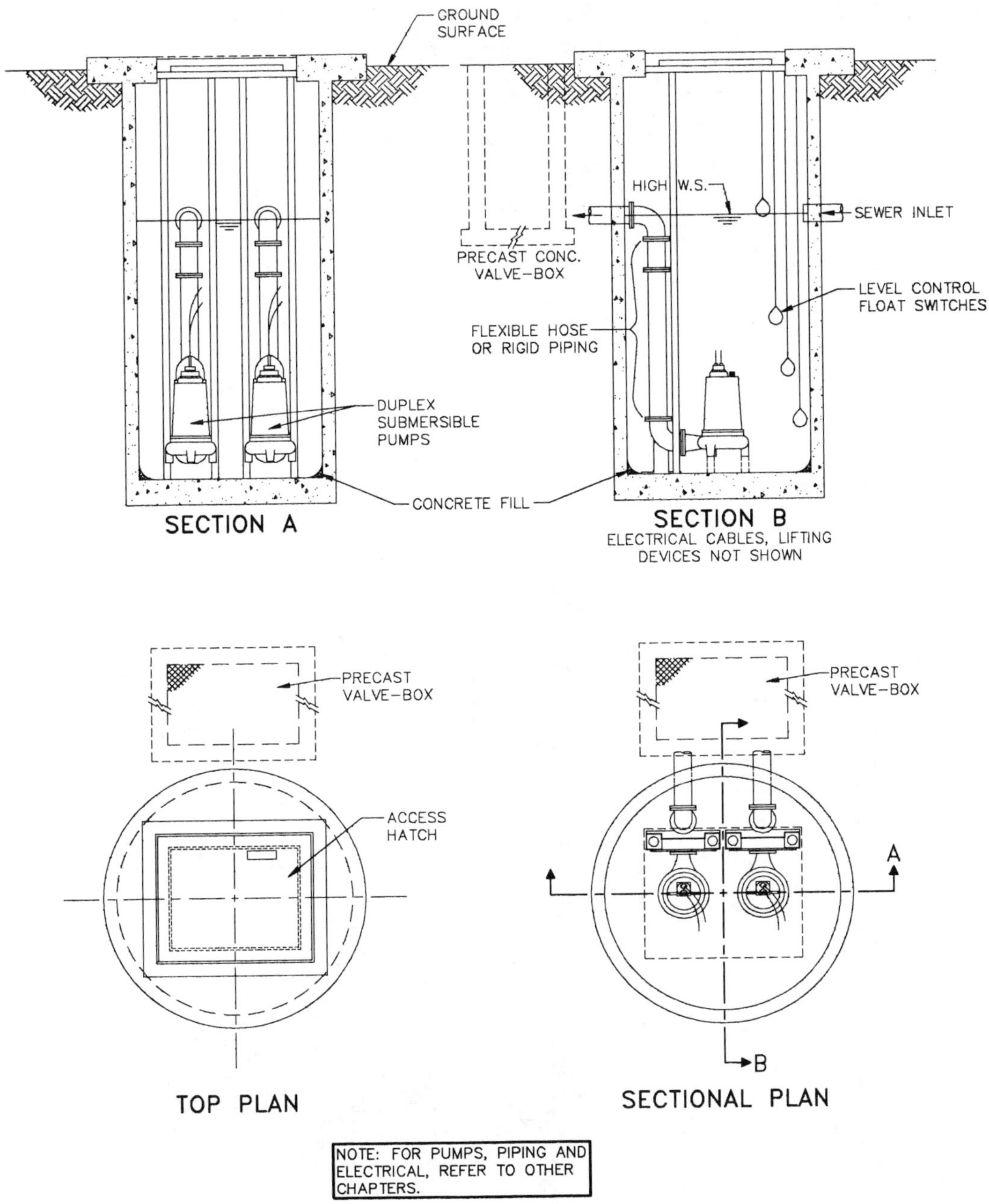

Figure 3.1 Package submersible pumping station configuration.

Pneumatic ejectors can be used instead of pumps, but they do not provide the flexibility to accommodate future growth, and generally need more maintenance than pumps. In the small flow range, use of pneumatic ejectors is least recommended because of inherent system energy inefficiency.

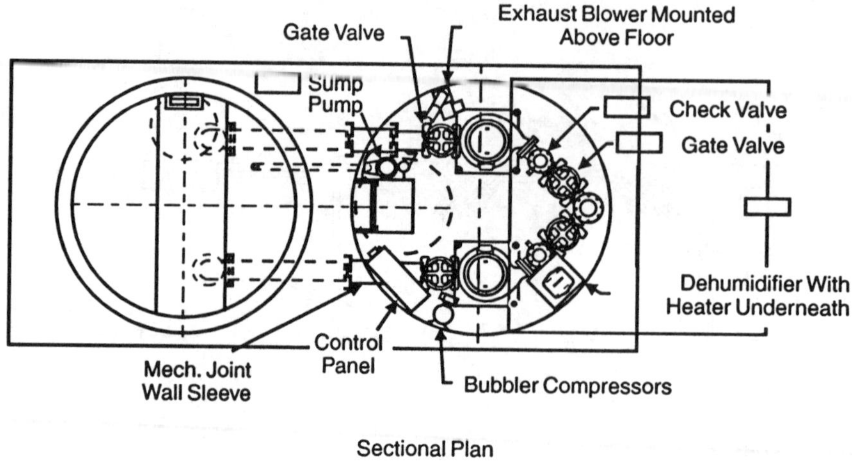

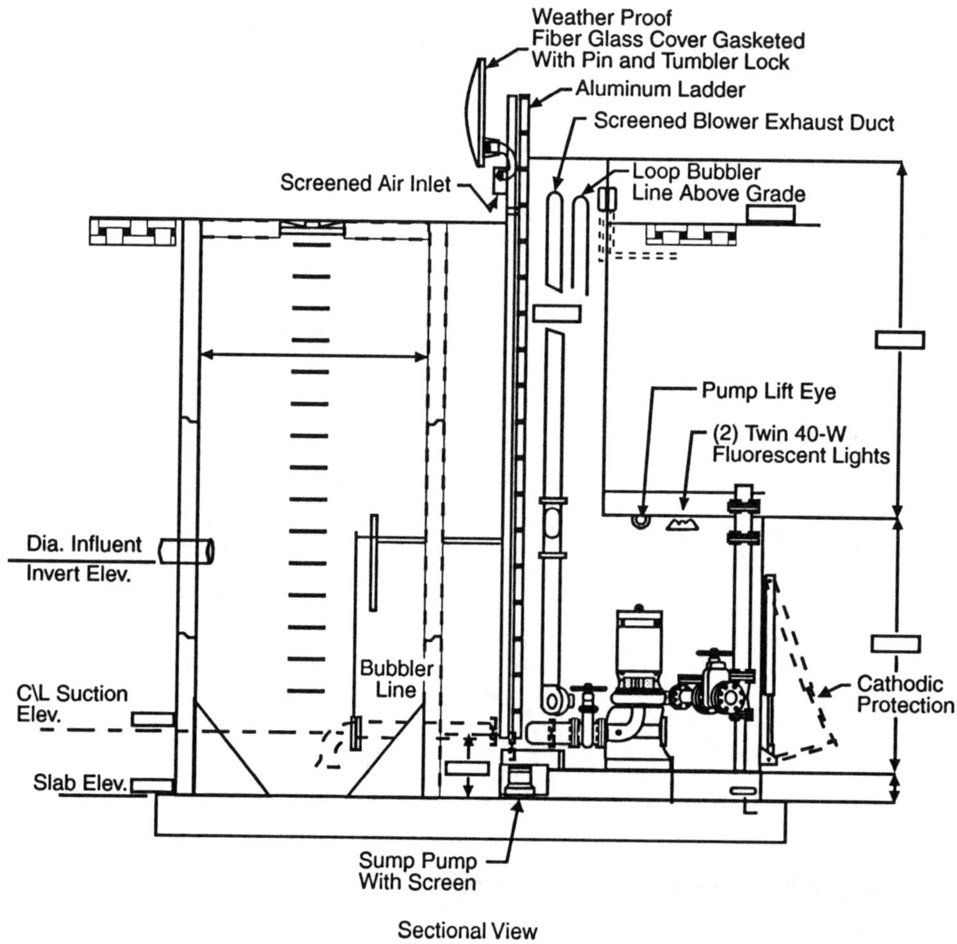

Figure 3.2 Package pumping station configuration (ft × 0.304 8 = m; in. × 25.4 = mm).

Small to medium stations handling 30 to 200 L/s (500 to 3 200 gpm) typically incorporate a minimum of two nonclog pumps, each having the capacity to match the maximum design flow, with the second pump providing 100% standby capacity. As the design capacity and pressure head requirements increase, a three-pump station should be studied, with the third pump providing 50% standby capacity. The three-pump station design provides added flexibility in meeting flow requirements. During an interim expansion stage, pumping capacity can be increased by replacing the impeller with the larger one, provided higher horsepower requirements are planned in the initial motor, motor starter, and electrical designs.

Though not to exclude prefabricated stations, the design engineer must consider access for maintenance and available built-in piping flexibility for future pump replacement or station expansion. Self-priming pumps should be limited to practical maximum suction lifts of approximately 6 m (20 ft). Use of self-priming pumps should be evaluated carefully because the need for a priming device adds another level of mechanical maintenance. To reduce electrical power surges for direct across-the-line starters for higher horsepower motors, soft start motor controllers and variable-speed drives (typically of the variable-frequency type) have applications in this flow range. The additional electrical panels required for such equipment may affect the station building size and arrangement.

Figure 3.3 illustrates an example of a built-in-place station for small to medium capacity. The discharge piping and valve arrangements are shown. Typically, design engineers prefer the check valve in the horizontal alignment to prevent jamming from solids. This is particularly important for stormwater pumping stations.

Air diffusers in the wet well may be considered as an optional design. They are used to freshen the flow stream (for odor control downstream, in lieu of the air injection to the force main sometimes used) and prevent the deposition of solids. Because aerating the flow stream can release volatile organic carbons and odor-causing agents to the air, to comply with air pollution control regulations the inclusion of air scrubbers should be studied, particularly if there is an industrial component in the flow stream. If preventing the deposition of solids is the only concern, the use of shaped and sloped wet wells is recommended.

A two-pump arrangement can be used in larger stations handling more than 200 L/s (3 200 gpm), but to deal with varying ranges of flow efficiently often requires three or more pumps. A combination of smaller pumps and one large pump will provide flexibility for low flow rates, especially during the start-up stage when flows can be low. Pump capacities should be selected to convey the maximum flow with the largest pump out of service. If variable-speed drives are to be installed, the drives might be sized based on larger future pumps (or a larger impeller) to save the cost of replacing them when pumping capacity is increased.

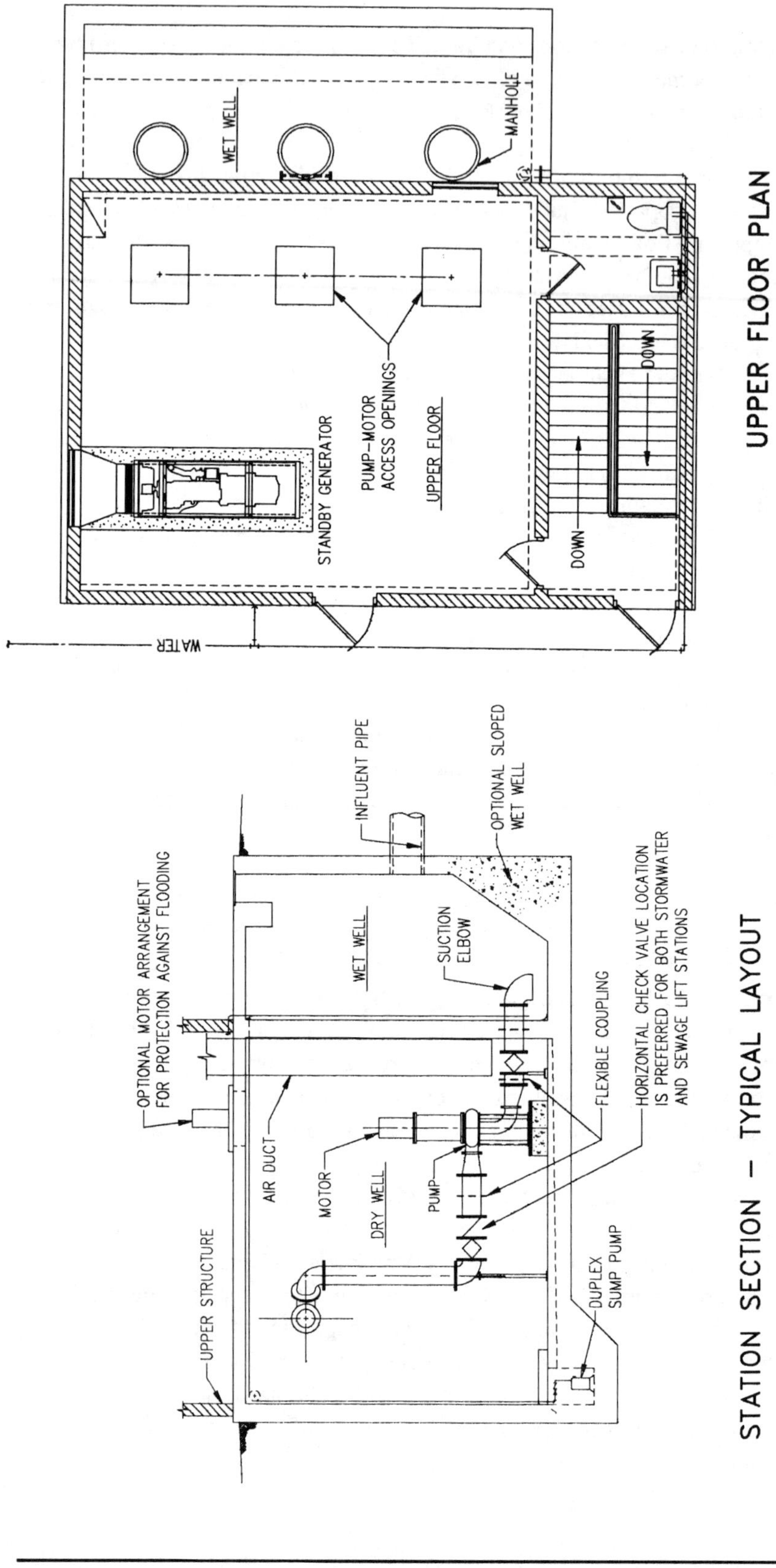

Figure 3.3 Small to medium pumping station configuration.

While prefabricated stations function adequately in certain medium flow range applications, a built-in-place station is the norm and provides greater flexibility and better satisfies site constraints and operation and maintenance requirements. Stormwater stations can use pump types of either dry pit, wet pit, or inclined screw pumps. The use of screw pumps typically is limited to lifts of approximately 8 m (25 ft). Figure 3.4 shows an arrangement for an inclined screw pump station.

Figures 3.5 and 3.6 show general layouts of medium to large built-in-place wastewater pumping stations. Figure 3.5 shows the plan of a station with a five-pump design (three existing pumps, two future pumps). The entire station hydraulic structure was built during the first stage because of extensive dewatering cost. Because of a wider range between initial and future design flow, a dual force main was considered to provide flexibility during initial and future pump selections. This condition is less likely for stormwater pumping stations.

Figure 3.5 shows an intermediate floor added to strengthen the hydraulic structure walls against hydrostatic pressures from high ground-water level. Air diffuser headers in the wet well provide odor control, and any pumping problem caused by air addition is prevented by installing a small-diameter recirculation line from each pump discharge to the wet well. Because of space and height of hydraulic structure, Figure 3.5 shows the check valve in vertical arrangement rather than the normal horizontal position. Groove-type couplings (for quick disassembly) for the valve and piping provide ease of maintenance.

Figure 3.6 shows a general layout of a pumping station designed to handle 11 000 L/s (250 mgd) of flow. Wet well configuration must be designed based on approach velocity and follow established standards such as those set by the Hydraulic Institute.

Figure 3.7 shows a wet-pit-type pump station for large [7.5-m^3/s (120 000-gpm)] stormwater pumping flows.

Stormwater pumping typically is used only when gravity drainage is not feasible, although there may be cases in which it is more economical to pump than to construct gravity conveyance systems. Storage in storm drains, wet wells, or ponds can reduce the pumping capacity required to handle peak runoff rates.

Several important considerations affect planning and site selection for pumping stations. Access necessary for safe operation, maintenance, and emergency functions must be available at all times. Hydraulic conditions will have primary importance in site selection, but site appearance and sound attenuation should also be assessed. Foundation investigations are essential (although they rarely reveal conditions unstable enough to merit rejection of a site); the effects of soil erosion and contaminated runoff on the pumping station influent and the rate of discharge and water quality of the effluent should

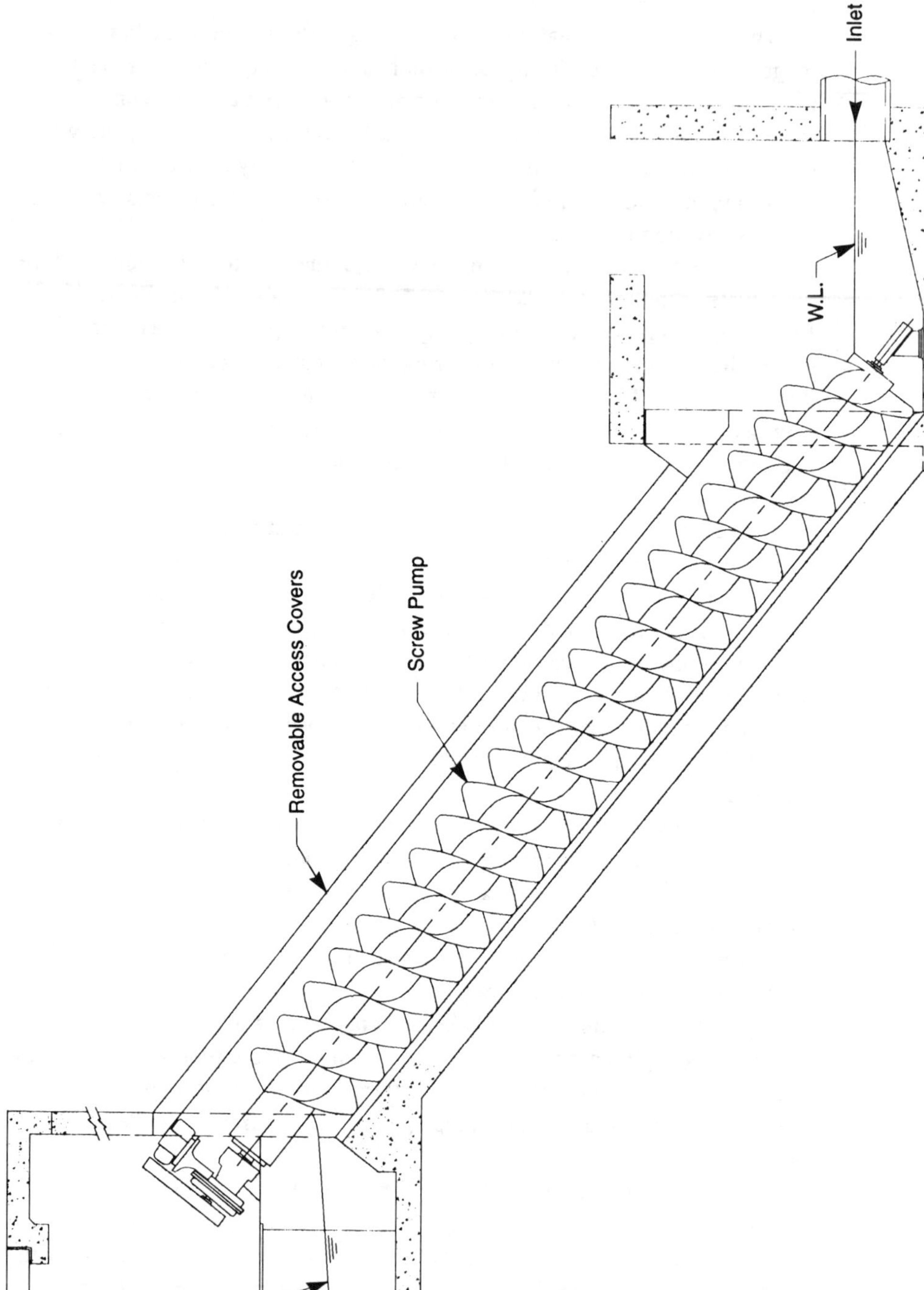

Figure 3.4 Screw pump arrangement.

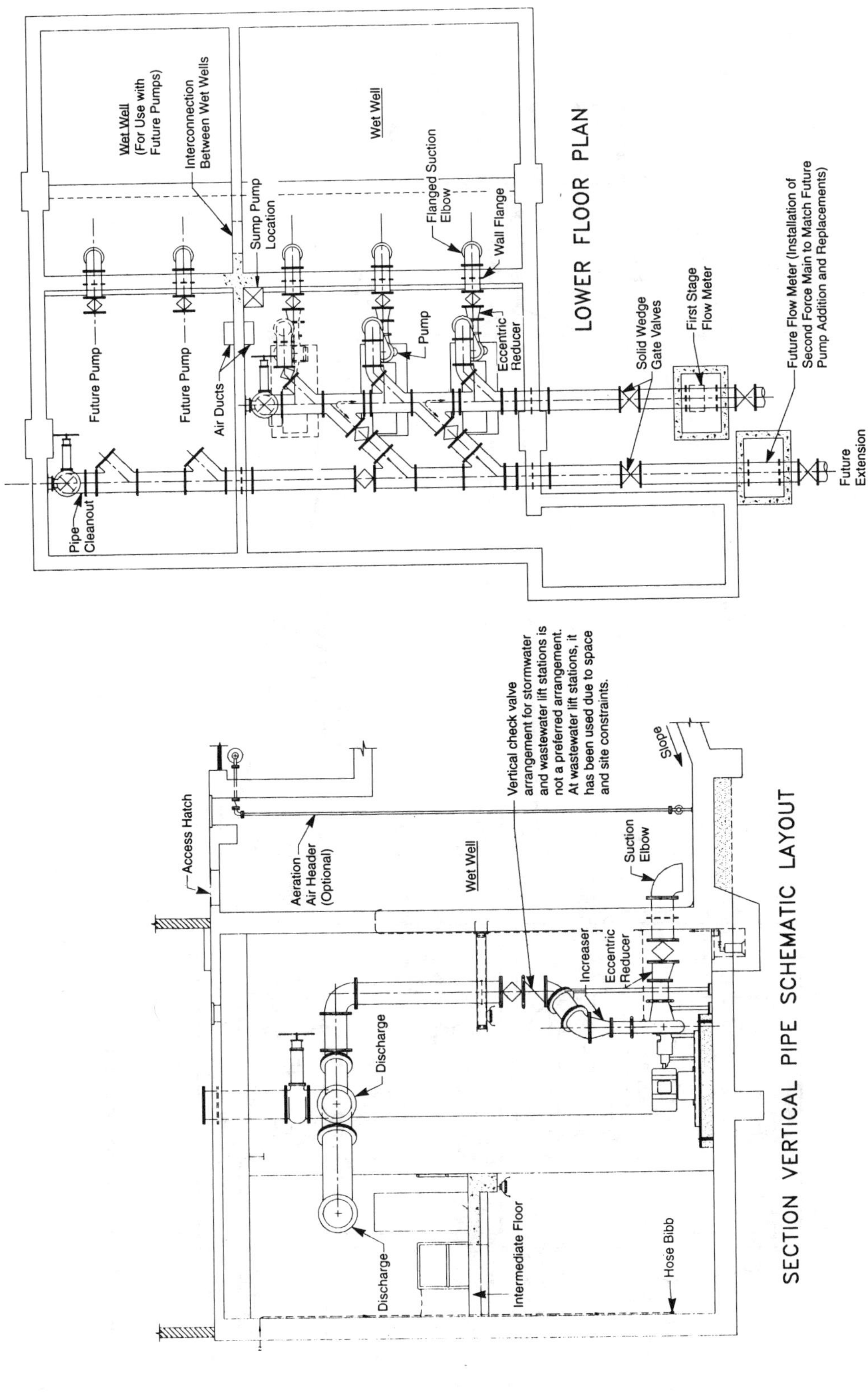

Figure 3.5 Medium to large pumping station configuration.

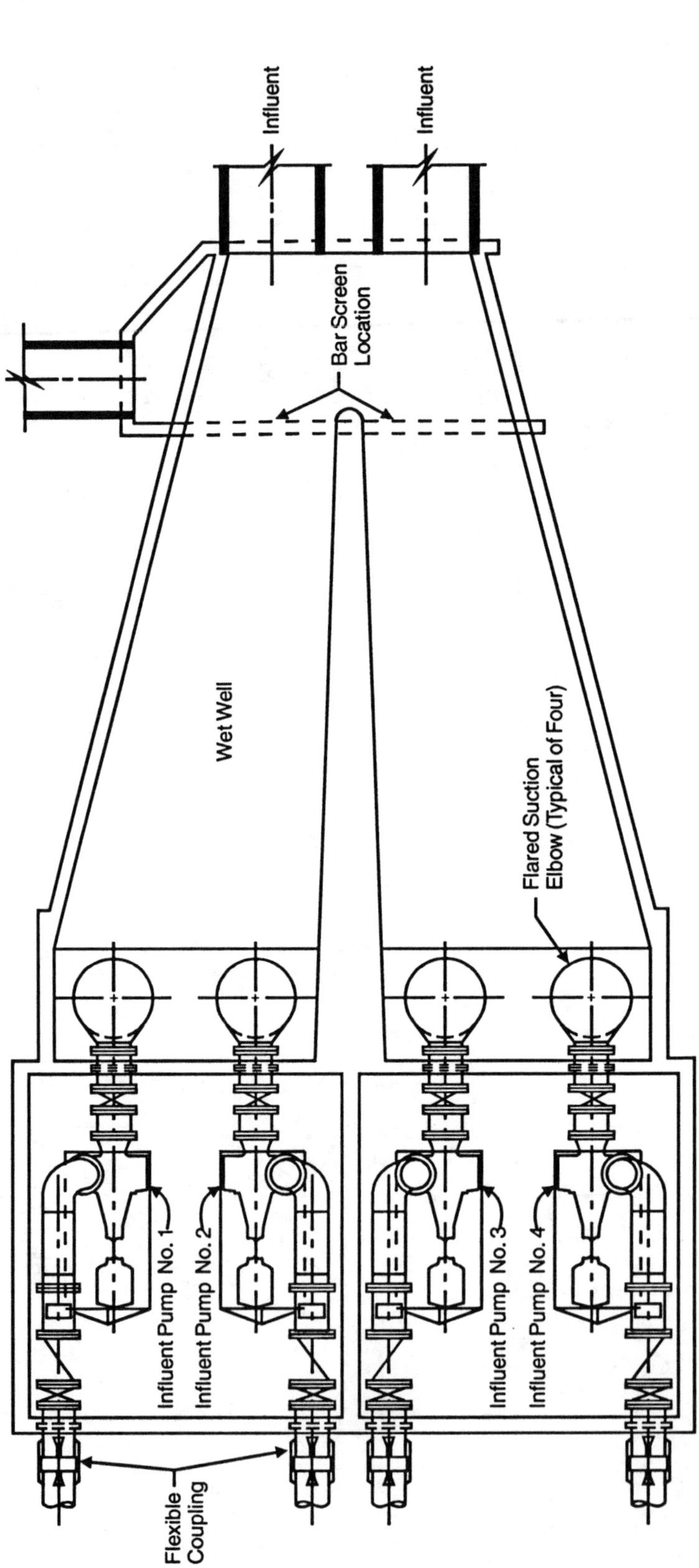

Figure 3.6 Large pumping station configuration.

be evaluated. Enough space must be provided in the area outside the station to accommodate parking and movements of large machinery.

A dependable energy source is essential. The primary source of electrical power for most stormwater pumping stations is a public utility. Underground service is preferred for safety and aesthetic reasons, and overhead lines into the station should be avoided, as they present potential safety hazards during large equipment operation.

Some features to evaluate when designing a pumping station depend on the size of the facility, while others relate to site conditions, economic considerations, and operating needs. Table 3.1 lists these features and rates the parameters of each in terms of high, medium, and low conditions.

Stormwater storage can reduce the required peak capacity of a pumping station. Selection of a peak pumping capacity is a trial and error process that considers the inflow hydrograph, available storage, and possible pump discharge rates.

Pumps commonly are used in storm drainage systems, though the designer is cautioned that they should be used only when gravity flow is not possible.

Stormwater pumping typically uses large pumps that operate at relatively low heads. The main types of pumps include

- Vertical (propeller and mixed flow);
- Submersible (vertical and horizontal);
- Centrifugal (horizontal nonclog);
- Screw; and
- Volute or angleflow (vertical).

All of these pumps typically are driven by electric motors, but diesel or gas engines sometimes are used for vertical, screw, or angleflow pumps.

Vertical Pumps. Vertical pumps, the type used most frequently in stormwater pumping stations, are available in many capacities and discharge heads. Single-stage propeller pumps are used for low heads; mixed-flow pumps are used for higher heads. Two-stage propeller pumps are available that will approximately double the head capacity of a single-stage pump. A typical vertical stormwater pump is shown in Figure 3.8.

Submersible Pumps. Until recently, submersible pumps, except at small stations, have been mainly used as accessory pumps to handle low (dry weather) flows, ground-water infiltration, and cleanup and pumpout of the pump pit after a storm. This has been changing as larger submersible pumps have become available, and large stormwater pumping stations can now be designed that rely solely on submersible pumps.

Table 3.1 Pump station design features and considerations (U.S. DOT, 1982).[a]

Feature or consideration	High condition	Medium condition	Low condition
Station design capacity	Maximum exceeding 300 cfs	Maximum between 100 and 300 cfs	Maximum less than 100 cfs
Station design head	More than 35 ft total dynamic head	Between 15 and 35 ft total dynamic head	Less than 15 ft total dynamic head
Storage upstream of pumps	For velocity reduction, settlement of solids, minimizing equipment	Used if available	Not required or available
Quality of pumped water	Turbid and sand-laden inflow	Moderate contamination	Minimal contamination
Inflow rate	Rapid increase	Normal hydrograph	Slow increase
Discharge conditions	Long rising outfall from each pump	Short free outfall from each pump	Limitation of discharge rate
Sump dewatering	Sump pump required	Vacuum truck preferred	No provision
Electric power reliability	Completely dependable; dual service	Very dependable; single service	Undependable; frequent outages
Natural gas/liquid petroleum gas desired as fuel	Completely dependable; dual service	Very dependable; good storage	Not readily available; supply unreliable
Station siting	Good access from frontage road or similar	Good access from highway	Poor access, alongside highway
Soil conditions	Rock	Hard, steep unshored cuts	Clay or soft soil
Foundation conditions	Acceptable bearing strata	Piling required for bearing	Extensive dewatering with piling required because of uplift
Aboveground structure	Large acceptable	Moderate preferred	Smallest possible desired
Structure visibility	Large structure acceptable	Modest structure desired	Minimum only acceptable
Initial cost	High capital cost acceptable	Moderate cost acceptable	Lowest cost mandatory
Maintenance capability	Excellent, with complex machinery	Reasonably good	Mediocre
Operating cost	High cost acceptable	Moderate budget desired	Lowest budget desired
Equipment handling devices (built-in)	Elaborate type considered essential	Single type acceptable	Minimum or none required
Equipment handling devices (mobile)	Use preferred for all requirements	Used to supplement built-in	Not required due to elaborate built-in

Table 3.1 Pump station design features and considerations (continued) (U.S. DOT, 1982).[a]

Feature or consideration	High condition	Medium condition	Low condition
Trash handling devices	Elaborate built-ins preferred	Simple built-in adequate	Vacuum trucks preferred
Prescreen inflow debris	—	—	—
Hazardous spill vulnerability	—	—	—
Epoxy coatings and lining of pumps	—	—	—
Grease lubrication for pumps	—	—	—
Steelwork galvanizing	—	—	—
Manifold to pressure discharge	—	—	—
Sediment and hydrocarbon removal from discharge	—	—	—
Emergency generator	—	—	—
Supervisory control (telemetering)	—	—	—

[a] cfs × 28.32 = L/s; ft × 0.304 8 = m.

Centrifugal Pumps. A special type of centrifugal pump is used in dry-pit stations. Designed to handle suspended solids and debris that pass the screens, these pumps are the end suction, nonclog type. Enclosed impellers and a handhole on the casing permit inspection and removal of foreign material. Small centrifugal pumps combined with upstream storage are well suited to small catchments.

Screw Pumps. Screw pumps are appropriate when pressurized transfer is not required. They typically are used in agricultural areas and offer a simple and reliable method of stormwater pumping when site conditions are suitable.

Volute or Angleflow Pumps. The volute or angleflow pump, also called a dry-pit angleflow pump or a single-suction centrifugal mixed-flow pump, usually is mounted in a vertical position, with the motor above the pump room operating floor and the pump as much as 7.5 m (25 ft) below. Vertically mounted volute or angleflow pumps have been a standard in wastewater pumping station installations for years. However, because of the complex accessories needed with these pumps, they are not recommended for stormwater pumping unless the level of operating head makes it necessary. For a more in-depth discussion of pumps and pumping systems, refer to Chapter 4.

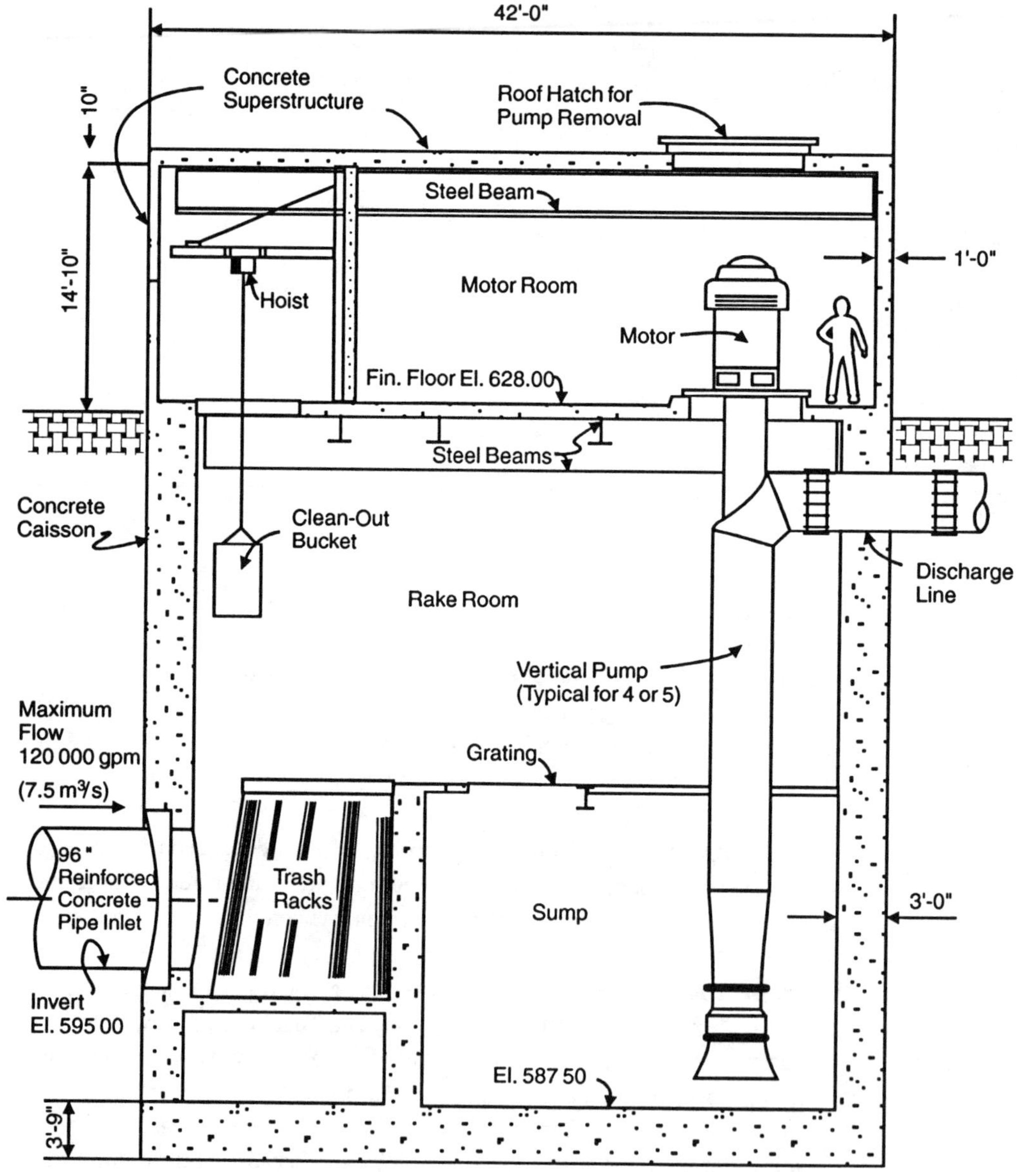

Figure 3.7 **Section through circular wet-pit-type pump station (dimensions and elevations shown are for a particular case and may be varied to suit) (gpm $\times$ 6.308 $\times$ 10^{-5} = m^3/s).**

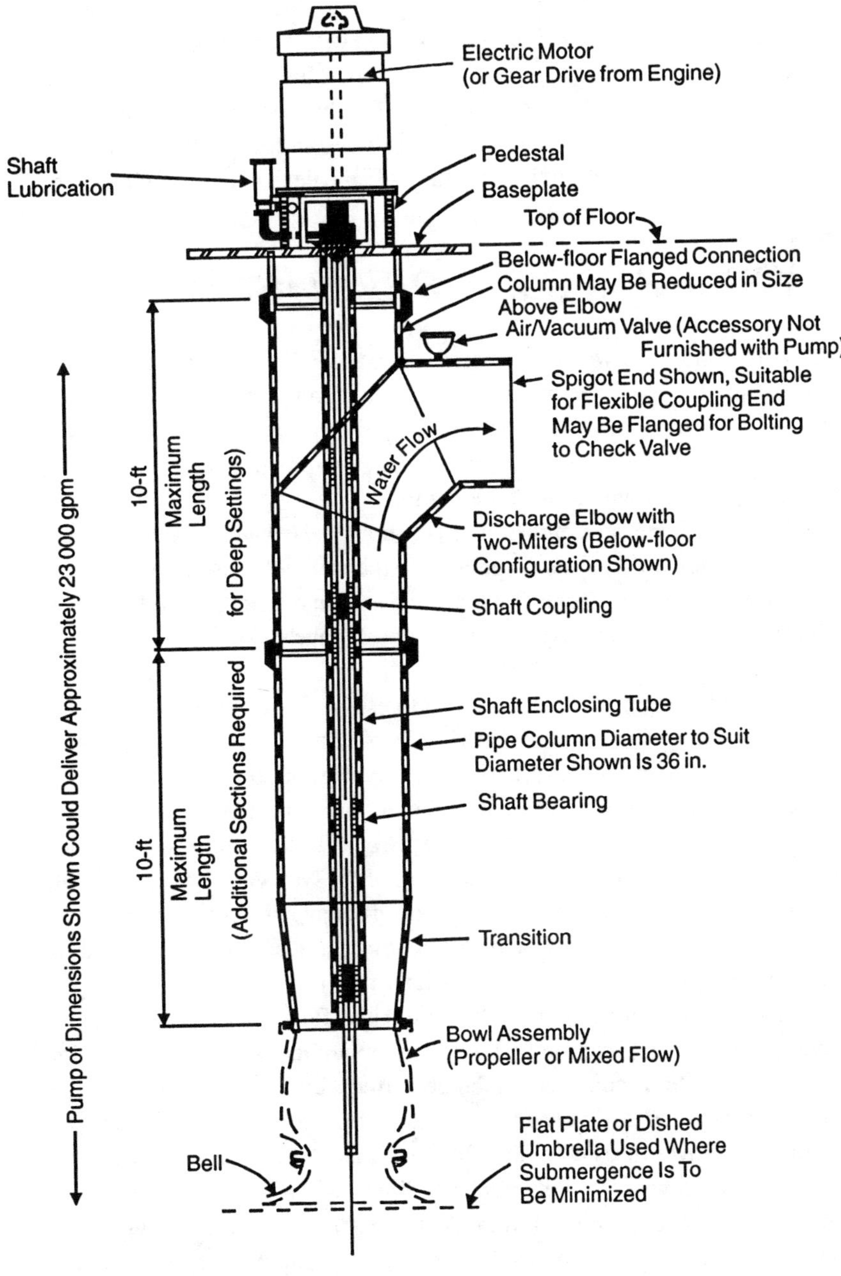

Figure 3.8 Typical vertical stormwater pump (oil-lubricated enclosed shaft shown) (gpm × 0.063 08 = L/s, ft × 0.304 8 = m, and in. × 25.4 = mm) (Federal Highway Administration, 1982).

For all pumping stations, the use of pumps with non-clog-type impellers is recommended. Screens or comminutors should be used where the size of solids in the flow stream exceeds the capability of the pumps to handle them.

It is a good practice for the design engineer to discuss new ideas and the special requirements of each pumping station with pump manufacturers and operation and maintenance personnel. Hydraulic Institute standards and publications such as those by the British Hydro-Mechanics Research Associations provide guidelines for dry well and wet well design for wet-pit pump installations. Simplicity and reliability should guide the decision-making process.

WET WELL CONSIDERATIONS

Wet wells store wastewater and stormwater (typically at smaller stations) before pumping, provide submergence over the pump suction, and provide a smooth transition of flow from the sewer to the pump suction. The wet well influent channel may also house appurtenant equipment (such as comminutors or bar racks) that protect the pump.

Design flow rate determination affects both wastewater and stormwater pumping station configurations, including the wet well. The selected design life of the station is also a key factor in the station configuration. Range of minimum and maximum daily flows and peak hourly flows should be evaluated carefully.

The design of wet wells is important to establish adequate hydraulic flow patterns toward pump suctions. Suction flow requirements for dry-pit pumps are less critical than for wet-pit, axial-flow pumps because the suction pipes to the pump can reduce prerotation and other potential problems.

The suction of dry well pumps is normally a flare fitting on a suction elbow in the wet well or on the vertical wall of the wet well. Concerns for these intakes include approaching velocity and sufficient submergence to prevent vortexing of air into the suction bell. Suction piping should be designed for the maximum pumping capacity of each pump.

The wet well configuration for wet-pit, axial-flow pumps is critical to ensure even flow to the suction bell of the pump without surface or submerged vortices. Items to be considered include currents created in the wet well by the incoming flow, flow past an inoperative pump to an operating pump, and submergence over the suction bell. In many installations, baffles to interrupt high-velocity flow and training walls to direct flow to a pump suction are required. For large or complex pumping stations, it is advisable to study the hydraulic flow patterns in a reduced scale hydraulic model of the wet well. Sometimes model studies for similar designs can be used.

It is important to reduce or eliminate free fall of water into the wet well to prevent entrained air from entering a pump suction. Entrained air forms a large bubble at the eye of the impeller that can cause vibration, reduce pump capacity, and even stop the pump from functioning by air binding it.

To allow air bubbles to rise to the surface, there must be sufficient distance from the point of free fall to the pump suction inlet. Inlet drop pipes or baffling help divert air from the pump suction. Also, a steeply sloping wet well floor helps separate the air from inflowing liquid.

In large systems, especially those with inlet screens, the high wet well level should be set at the highest possible level without surcharging the influent sewer. This reduces free fall at low wet well levels.

The required storage volume in the wet well depends on the method of pumping used. Pumping stations with on–off control of the pumps require sufficient storage in the wet well to prevent the pump motors from short cycling and overheating. For dry-pit motors, a maximum of six starts per hour is recommended for motors 20 kW (25 hp) and smaller, four starts per hour for motors between 25 and 75 kW (33 and 100 hp), and two starts per hour for motors between 100 and 200 kW (135 and 270 hp). Larger, costlier motors that allow more starts per hour are available, but these motors are slightly less efficient than conventional motors. The motor manufacturer should be consulted on the allowable starting frequency of such motors.

Submersible pumps have motors designed to operate in liquid, and most such motors are designed for up to 10 starts per hour whether the motor is submerged or not. Some larger motors are cooled by circulating some of the pumped fluid through a cooling jacket. The pump manufacturer should be consulted for allowable frequency of starts, especially if the motor is not submerged.

Elapsed time between starts is a function of the pump discharge rate and the inflow to the wet well. In the case of a multispeed pump or multiple pumps, it is determined by differences among step flow rates when the inflow is between the two steps. The time between starts is at a minimum when the inflow is 50% the pump rate half way between two steps. The general formula for computing the required wet well storage is

$$V = tQ/4 \tag{1}$$

Where

V	=	storage volume, m^3 (gal);
t	=	required time between starts, minutes; and
Q	=	pump discharge capacity, or the difference in flow rate between steps, m^3/s (gpm).

Many public agencies specify a maximum storage time in a wet well to reduce the potential for odors and septic wastewater. A maximum storage retention time of 30 minutes at average flow is a common requirement. Shorter retention time can conflict with requirements to prevent short cycling. In such cases, multispeed (including variable-frequency drives) pumps or multiple pumps should be considered.

Variable-speed pumps require little storage time because they match the incoming flow rate, thus eliminating frequent starts and stops. Most controls are set to maintain maximum speed at high wet well level and minimum speed at low wet well level. The vertical distance between the two levels and the resulting storage in the wet well must be sufficient to allow stable operation of the speed controls in the event of sudden changes in incoming flow.

Wet wells should be configured to prevent accumulation of settleable solids that could occur independent of the pump cycling time. For large pumping stations, compartmentalization for the purpose of guiding flow to each pump suction is an option.

Sewer gas generated in the sewer and volatile material spilled into the sewer are transported and exposed to the atmosphere in the wet well. Therefore, all equipment (particularly electrical motors) in the wet well must be of nonvolatile, explosion-proof construction.

If the station has bar racks or screens, another consideration is the control of flow velocity through the bars. Velocity through the bars is a function of flow rate and wetted area and can be controlled by the wet well levels. The velocity through the clear opening between bars should be between a maximum of 1.2 m/s (4 ft/sec) and a minimum of 0.4 m/s (1.2 ft/sec). At high wet well levels (high speed) both the flow rate and the wetted area are at a maximum. At low wet well levels (minimum speed), the flow rate is at a minimum, but the wetted area is also at a minimum.

AESTHETIC AND SAFETY CONSIDERATIONS

Aesthetic considerations can provide pleasant surroundings for the public and a comfortable atmosphere for station operation and maintenance personnel. Preliminary stages of pumping station design should evaluate numerous structural, architectural, practical, and safety constraints, including the following:

- Building material and configuration compatible with neighboring developments;
- Adequate ventilation for workers
  - Control of ventilation fans could be linked to entrance door open and close positions;
  - Continuous or timer-controlled air changes may be required to meet local safety regulations;
  - Separate ventilation for dry and wet well;

- Provisions for emergency eye wash and chemical leak alarm linkage with local fire station (if chemicals such as hydrogen peroxide and chlorine are used for odor control);
- Building construction of vandalproof material;
- Gas migration barrier between wet well and dry well [interior wet well coating (to guard against corrosion) typically offers this protection];
- Effective odor mitigation system to avoid offending neighbors;
- Unobstructed sight paths for neighboring properties;
- Potential for flooding from backflows because of high water level at the station wet well;
- Site drainage and erosion control integrated with good landscaping design;
- Choice of fencing to satisfy local zoning and historical regulations;
- Use of parapet walls to hide certain equipment [such as heating, ventilating, and air-conditioning (HVAC) ducting and generator muffler exhaust piping];
- Need for an on-site restroom;
- Sloped dry well floors to avoid accumulation of water;
- Noise control through use of acoustical materials at emergency generators and other sensitive locations;
- Exterior lighting consistent with that of neighbors; and
- Adequate parking and easy access for delivery of chemicals, maintenance of equipment, service or removal of pumps, motors, drives, and auxiliary equipment.

ARCHITECTURAL DESIGN CONSIDERATIONS

Station arrangements, building layouts, and safety features must satisfy the requirements of local, state, and national codes (such as those of the Occupational Safety and Health Administration). The abovegrade structure is designed to provide access to the lower levels of the pumping facility containing the electrical controls and ancillary equipment; the station design itself must be compatible with local zoning and site approval criteria. Factors include plans for lighting (discussed in more detail in Chapter 6), plumbing, and HVAC systems.

In residential and commercial areas, there should be adequate distance between buildings, if possible. Concrete masonry blocks (plain face or split face) and bricks are suitable building materials, as is stucco construction (wood or steel stud framing). The upper structure should have the appearance of a residential building. Windows (with steel wire-reinforced glazing for protection against vandalism) and false windows can be integrated with the building. Using the right approach, each pumping station can be pleasing and acceptable to the neighborhood property owners.

Adequate lighting with convenient switch locations should be provided for all stations. Skylights to enhance natural lighting are a possibility. Shadows and glare should be avoided near machinery and at floor openings. Deep dry wells should have multilevel light fixtures located such that maintenance will not be a safety concern. A motion-actuated exterior lighting system is an option.

To ensure adequate building drainage, designs should include a minimum lower floor slope of 10 mm/m (1/8 in./ft), and floor gutters are recommended for areas of chronic wetness. Floor drains should be provided in restrooms, adjacent to pumps, and adjacent to generator oil drain outlets (to collect accidental spills during maintenance). An exterior oil–water separator sump should be considered, and sump pumps actuated by level switches should be used to transfer the drainage from floor sump to the wet well. Check valves should be installed in each sump pump discharge line to avoid the backflow of wastewater to the sump. The main sump pump discharge piping should connect to the wet well at a point higher than the high water alarm level.

The station site should be graded, landscaped, and designed with properly vented drainage to conform to the surrounding properties and keep them free of drainage overflow.

For sensitive electronic control equipment, localized HVAC cooling systems can be used to mitigate potential damage from heat generated by large motors and emergency generators. Proper separation of HVAC ducting and electrical panels helps prevent interference; and, to screen rooftop HVAC equipment, exterior treatment such as using a parapet is an option.

STRUCTURAL DESIGN CONSIDERATIONS

Station hydraulic structures (substructures) can either be custom designed or preengineered (prefabricated package pumping stations). For small stations, a precast or cast-in-place concrete manhole [a minimum of 1.75 m in diameter (5 ft)] could serve both as wet well and submerged pumps housing.

The depth of incoming sewers or storm drainage channels, and sometimes foundation conditions, will determine the depth of the station substructure below the ground level. Surface conditions, environment, and necessary elevation to avoid flooding should determine the level of the operating floor, the type of superstructure, and the required exterior finish and protection.

The following criteria must be considered in designing the lower structures:

- All hydraulic structures must be designed to withstand uplift forces from high ground-water conditions and earthquakes.
- For stations more than 9 m (30 ft) deep, an intermediate (mezzanine) floor can be added to increase structural rigidity and provide support for drive shaft bearings (if motor and drive unit is placed at the upper floor) and lighting fixtures. This extra floor space could also be used for storage.
- For excavation, backfill, sump dewatering, and foundation requirements, reference the soil profile investigation. For deep station construction and in areas of high ground water, bidding documents should not be completed until sump pump and well point dewatering operations are approved by geologists, hydrogeologists, geotechnical engineers, and dewatering subcontractors. The engineer and client should determine if the contract is to provide specific excavation support system details or if the documents are intended to include only design criteria related to the development of an excavation support structure. To secure the lowest bid price, the client should allow specific design of the excavation support system to be the responsibility of the contractor.
- All entrances and unsealable openings in the station should be at elevations above the flood protection elevation.
- Station substructures typically should be made of reinforced concrete. For small, prefabricated stations, steel plate construction and fiber glass are acceptable. American Concrete Institute and American Institute of Steel Construction standards should be followed, in addition to the Uniform Building Code and applicable local codes.
- Designs must provide for practical ways to remove equipment from the pumping station by including access openings and hatches, whether it be a separate access hatch per pump or a single opening (refer to Figures 3.3 and 3.5). Skylights can serve as access hatches in large, deep, vertical pump stations.
- For small stations, A-frames and dollies are needed to lift pumps, motors, and gears. For large stations, a monorail or a traveling bridge crane can be used to lift heavy items, though such hoisting equipment needs adequate space in which to operate.
- All entrances should have handrails and gratings, with nonslip treads standard for all stairways.
- Access hatches and safety ladders or stairs should allow for proper inspection and maintenance of both dry and wet wells. Ladders must be provided with safety cage or harness arrangements, with stainless steel or aluminum ladders recommended for the wet well.
- Because wastewater pumping stations continuously are exposed to corrosive environments, wet wells can be equipped with a chemical feed system for odor control. Wet well walls and the bottoms of top

slabs should be coated (with coal tar epoxy or other proven coating systems) or lined with polyvinyl chloride, though this latter alternative is less common. Wet wells must be provided with positive forced ventilation, and their floors typically are sloped to guide the solids toward the pump suction elbows.

- Adequate ventilation is essential to protect the dry well from moisture, especially in the presence of high ground water.
- Consideration should be given to the use of package stations, which are factory-fabricated enclosures with special valves and fittings designed for compactness. The enclosure typically is constructed of steel with various options of coating and cathodic corrosion protection systems. Prefabricated reinforced concrete stations are also available, and some manufacturers offer fiber-glass enclosures. Package stations have the advantages of low initial cost and rapid installation, and can be relocated. However, the cost of rehabilitating and retrofitting piping inside pumping stations should be evaluated carefully. Disadvantages of package stations include a limited operating life because of corrosion, restricted working space, and inflexible piping for pump replacement and station expansion.
- Structural design requirements for built-in-place concrete stations should also apply to package stations, which require a certain amount of engineering design analysis, such as design of a separate wet well and a foundation slab. Specific tie-down designs could be required for lightweight packaged stations at high ground-water locations.
- Access to the inside of a wet well for removal of sand, grit, grease, and large objects should be considered. The addition of bar screens, comminutors, and other pretreatment devices affects wet well design.

SUGGESTED READINGS

Hydraulic Institute Standards for Centrifugal, Rotary & Reciprocating Pumps (1975). 14th Ed., Hydraul. Inst., Cleveland, Ohio.

Odor Control & Mitigation (1987). Calif. Assoc. of Sanit. Agencies, Sacramento.

Odor Control for Wastewater Facilities (1979). Manual of Practice No. 22, Water Pollut. Control Fed., Washington, D.C.

Chapter 4
Pumping Systems

Pumps and pumping systems play a vital role in wastewater and stormwater systems. Therefore, it is imperative that design engineers produce well-engineered pumping stations that operate reliably and efficiently.

This chapter provides design criteria for wastewater and stormwater pumping stations that are located in collection systems. These vary from small, simple stations in remote areas of collection systems to large, complex pumping stations that discharge all or a major portion of a collection system's wastewater to a treatment plant.

Also discussed in this chapter are characteristics of pumps and drives, fundamentals of system hydraulics, the use of system head-capacity curves and the selection of equipment for reliable, efficient pumping systems to meet the needs of the system head-capacity curve.

Pumping station design must recognize and blend many requirements into an integrated scheme. Some of these requirements are discussed further in other chapters of this manual.

Several chapters in *Design of Municipal Wastewater Treatment Plants* (1992) discuss pumping stations and systems at wastewater treatment plants. These include influent and effluent pumping stations and sludge, scum, grit, chemical solution, and utility water pumping systems.

Wastewater and Stormwater Pumping Stations

There are similarities and differences between wastewater and stormwater pumping stations. Both receive flow from a drainage system and discharge the flow. An obvious difference is that wastewater pumping stations are in continuous operation, whereas a stormwater pumping station may be idle for considerable periods of time between storm events.

An important factor in the design of a pumping station is the type of sewer system that delivers flow to the station. The three types of sewer systems are sanitary, storm, and combined sewers. A sanitary sewer carries infiltration and inflow but does not carry stormwater. Sanitary wastewater typically does not contain large solids [more than 75 mm (3 in.) in diameter] or large quantities of grit but can contain rags and stringy material that do clog pumps.

Storm sewers collect surface runoff, ground-water infiltration, and inflow. Stormwater flow typically contains large solids, stringy material, and large quantities of grit.

Combined sewers carry sanitary wastewater during dry weather periods but carry both sanitary wastewater and stormwater during wet weather periods.

Both wastewater and stormwater pumps often require screens or other protective devices to remove solids or stringy material that might clog pumps. Wastewater pumps, especially those discharging to a treatment plant, often remove the maximum quantity of material and reduce the load on the treatment process by having screens with only small openings between bars. Stormwater pumping stations, however, normally have large, clear openings between bars that protect the pump, yet reduce the quantity of screenings that must be handled at the station. Additional information on screens and comminutors is presented in Chapter 8.

The end suction centrifugal pump is the most frequently used type in wastewater pumping stations. Vertical pumps are the common type for stormwater applications. However, stormwater pumping stations generally have higher capacities at lower heads than wastewater pumping stations and often are only lift stations that receive flow from an interceptor and discharge to an open channel. Therefore, Archimedes screw pumps and axial-flow, wet-pit pumps are common in stormwater pumping stations.

Motor-driven pumps are more common than engine-driven pumps in both wastewater and stormwater pumping stations. Stormwater stations frequently are idle for extended periods of time but, when operating, have high power draw. This duty cycle may favor the use of engine-driven pumps. In selecting the type of engine drive, the higher operating and maintenance costs of the drive should be compared to the demand and energy charges for electrical power.

Figure 4.1 is typical of a medium-sized pumping station. The station houses four pumps driven by shafting, with the motors mounted on the floor above the pumps.

Figure 4.1 Medium-sized wastewater pumping station.

PUMPING SYSTEM HYDRAULICS AND DEFINITIONS

Pumps are used to transport water from one point to another. To accomplish this, energy must be expended. The amount of energy expended depends on the quantity (flow rate), resistance to flow (the head the pump must overcome), and efficiency of the pump and drives. Terms associated with pumping systems include capacity, head, and pump efficiency and power requirements.

CAPACITY. Capacity is the volume of liquid pumped in a unit of time. The capacity typically is expressed as litres per second (L/s) or cubic metres per second (m^3/s) [gallons per minute (gpm) or million gallons per day (mgd)].

HEAD. A pressure term, head is the elevation of a liquid surface above or below a reference datum elevation. For instance, if an open-ended tube were extended vertically from a pipe under pressure, the head would be the vertical distance from the centerline of the pipe to the free water surface in the tube. Head typically is expressed in terms of metres of fluid (feet of fluid).

Head can refer to both pump head and system head. Pump head is the height the pump can raise the liquid being pumped at a given flow rate. System head is the height to which the liquid must be raised to overcome the resistance of a pipe system at a given flow rate.

Many types of head are included in the analysis of a pumping system, as illustrated in Figure 4.2. The datum elevation typically is taken as the centerline of the pump. Heads above datum are taken as positive heads and heads below datum are negative.

Static head (H_{stat}) is the difference in elevation between the free liquid surface on the discharge side of the pump and the free liquid surface on the suction side of the pump. Therefore, H_{stat} is the algebraic sum of the discharge static head (h_{sd}) and the suction static head (h_{ss}). Suction and discharge static heads are the difference in elevation between the suction, or discharge liquid surface, and the centerline of the pump.

Velocity head is the kinetic energy contained in a fluid in motion and is given by the following formula:

$$H_v = \frac{V^2}{2g} \tag{1}$$

Where

H_v	=	velocity head, m (ft);
V	=	velocity of fluid, m/s (ft/sec); and
g	=	acceleration because of gravity, 4.81 m/s^2 (32.2 ft/sec^2).

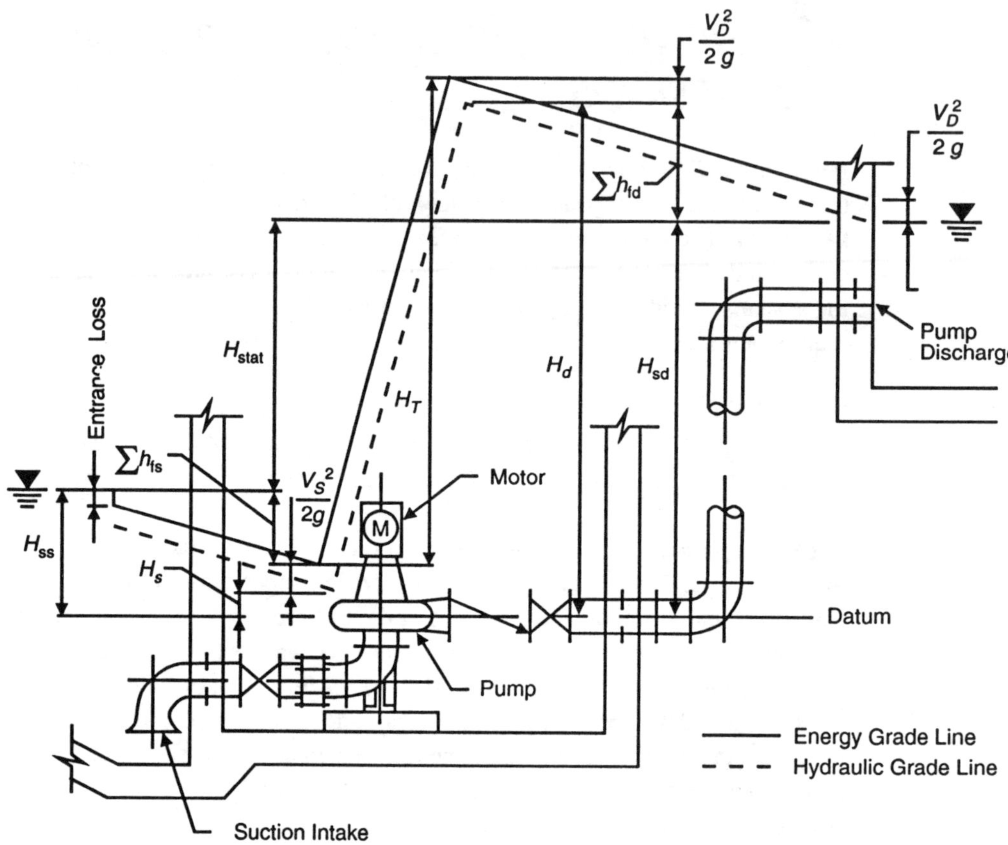

Figure 4.2 Definition sketch for pump heads: $V_S^2/2g$ = suction velocity head; $V_D^2/2g$ = discharge velocity head; H_{ss} = suction static head; H_{sd} = discharge static head; Σh_{fs} = total suction friction losses; Σh_{fd} = total discharge friction losses; H_s = suction head; H_d = discharge head; H_{stat} = net static head; and H_T = total pump head.

Because the velocity heads in both the suction and discharge piping are internal to the piping system; the only velocity head loss not recoverable in the system is the discharge velocity head where the pipe discharges to an open body of water.

Friction head (H_f) in a piping system is the head of liquid required to overcome the resistance to flow caused by the drag of the pipe wall on the fluid. As with static head, the friction head can be separated into suction friction losses (h_{fs}) and discharge friction losses (h_{fd}).

Minor head losses (h_m) are head losses required to overcome the resistance to flow through fittings and valves. Minor losses can be added to friction losses because both are caused by the resistance to flow in a pipe system. Figure 4.2 shows friction and minor losses in the suction combined, shown as Σh_{fs} and, in the discharge, as Σh_{fd}.

Total head is the algebraic total of the suction and discharge heads as given by the following equation:

$$H_T = H_d - H_s \tag{2}$$

and

$$H_d = H_{sd} + \Sigma h_{fd} + H_{vd} \tag{3}$$

$$H_s = H_{ss} - \Sigma h_{fs} \tag{4}$$

Where

H_T	=	total dynamic head, m (ft);
$H_d(H_s)$	=	total discharge (suction) head measured at the pump nozzle with static head based on pump centerline, m (ft);
$H_{sd}(H_{ss})$	=	discharge (suction) static head, m (ft);
Σh_{fd} (Σh_{fs})	=	the total of the discharge (suction) friction and minor losses, m (ft); and
H_{vd}	=	discharge velocity head at the end of the force main, m (ft).

If the suction and discharge static heads are combined, equation 3 can be rearranged as

$$H_T = H_{stat} + \Sigma h_{fd} + \Sigma h_{fs} + H_{vd} \tag{5}$$

Where

H_{stat} = total static head, m (ft).

PUMP EFFICIENCY AND POWER REQUIREMENTS. The amount of energy required for pump operation is proportional to the rate of flow, the head required to overcome system resistance, and the efficiency of the pump. This energy, or power, is measured in kilowatts (kW) or horsepower (hp).

In the metric system, when pumping water, the pump input power is

$$P_s = \frac{QH_T}{0.102E_p} \tag{6}$$

Where

P_s	=	input shaft power, kW;
Q	=	flow rate, m^3/s;
H_T	=	total head, m; and
E_p	=	pump efficiency, dimensionless.

In English units, when pumping water and the flow is given in gpm, the pump input brake horsepower is

$$P_s = \frac{QH_T}{3960E_p} \quad (7)$$

Where

P_s	=	input shaft power, hp;
Q	=	flow rate, gpm;
H_T	=	total head, ft; and
E_p	=	pump efficiency, dimensionless.

It should be noted that the above is input energy to the pump. To compute the total power of the pump and drive, the pump input power must be divided by the efficiency of the driver.

PUMPS

Pumps typically are categorized as kinetic energy pumps or positive displacement pumps. Kinetic energy pumps produce high-velocity kinetic energy in the pump that is converted to pressure. Positive displacement pumps produce pressure by applying force directly to the fluid.

Kinetic energy pumps include centrifugal, peripheral, and special pumps. Centrifugal pumps are the most common type of wastewater pumps and can discharge large volumes of water and pass the solids normally found in wastewater or stormwater. Peripheral turbine-type pumps and special pumps (such as hydraulic ram or jet pumps) are not used in wastewater service because they require close clearances and cannot pass wastewater solids.

Positive displacement pumps include reciprocating piston and plunger, rotary (progressive cavity, gear and vane, peristaltic, and screw pumps), and blow case pumps. Many positive displacement pumps are used for pumping sludge and scum at treatment plants but are not common in wastewater and stormwater pumping stations because of their limited capacities and inability to pass large solids.

The two types of positive displacement pumps that are used in wastewater and stormwater pumping stations are the Archimedes screw pump and the pneumatic ejector. The Archimedes screw pump is a lift pump that can discharge large quantities of flow and the pneumatic ejector is used to discharge small flows that contain wastewater solids.

CENTRIFUGAL PUMPS

The centrifugal pump is the most common type found in wastewater pumping stations. Centrifugal pumps consist of two main parts, a rotating impeller and the stationary pump casing. The impeller is driven by the pump shaft and transfers energy to the fluid being pumped. This energy produces flow and head. The casing directs the flow to the pump discharge and converts some of the flow velocity to head.

PUMP CLASSIFICATION. Centrifugal pumps are classified as radial-flow, mixed-flow, and axial-flow pumps, where the type of pump describes the shape of the impeller and the discharge flow pattern in relation to the pump shaft. Flow enters the impeller parallel to the pump shaft. Discharge from a radial-flow impeller is perpendicular to the pump shaft, discharge from an axial-flow pump is parallel to the pump shaft, and discharge from a mixed-flow impeller is at an angle to the shaft.

The shape of the casing will also depend on the type of pump. Radial and some mixed-flow pumps have volute-type casings that collect the flow as it leaves the impeller and guides it to the discharge nozzle, which is perpendicular to the pump shaft. Axial-flow pumps and some mixed-flow pumps are fitted with diffusers that straighten the flow as it leaves the impeller so that the flow is parallel to the shaft.

The shape of the impeller determines the flow pattern. A type number, known as specific speed, has been developed that characterizes pump type and performance. Specific speed is based on pump performance at its best efficiency point (BEP) and pumps of the same specific speed will have similar head-capacity curves. The formula for specific speed is

$$N_s = \frac{NQ^{0.5}}{{H_T}^{0.75}} \tag{8}$$

Where

N_s	=	specific speed;
N	=	pump rotative speed, r/min (rpm);
Q	=	flow rate at BEP, m^3/s (gpm); and
H_T	=	total head at BEP, m (ft).

The specific speed in metric units can be multiplied by 51.6 to be converted to English units. When calculating the specific speed for double suction pumps, only the flow through one side of the impeller, or 50% of the total flow, should be used. However, double-suction pumps are not typically used in wastewater or stormwater service.

Specific speed has value because it indicates the type of pump and is useful in predicting the performance of a pump. The Hydraulic Institute Standards have tables and curves that predict, based on pump specific speed, design and cavitation characteristics and abnormal operation of pumps.

Radial-flow pumps have specific speeds that range from 10 to 80 (500 to 4 000 in English units), mixed-flow pumps have specific speeds from 60 to 175 (3 000 to 9 000 in English units), and axial-flow pumps have specific speeds of more than 175 (9 000 in English units).

Figures 4.3 and 4.4 show the approximate shape of the impeller and the expected efficiencies for various-capacity pumps as the pump specific speed varies. Figure 4.3 is in metric units and Figure 4.4 is in English units. Mixed-flow impellers in the specific speed range of 60 to 125 (3 000 to 6 500 in English units) are available with volute-type casings with discharge perpendicular to the shaft and are known as mixed-flow volute pumps. Mixed-flow impellers in the specific speed range of 75 to 175 (4 000 to 9 000 in English

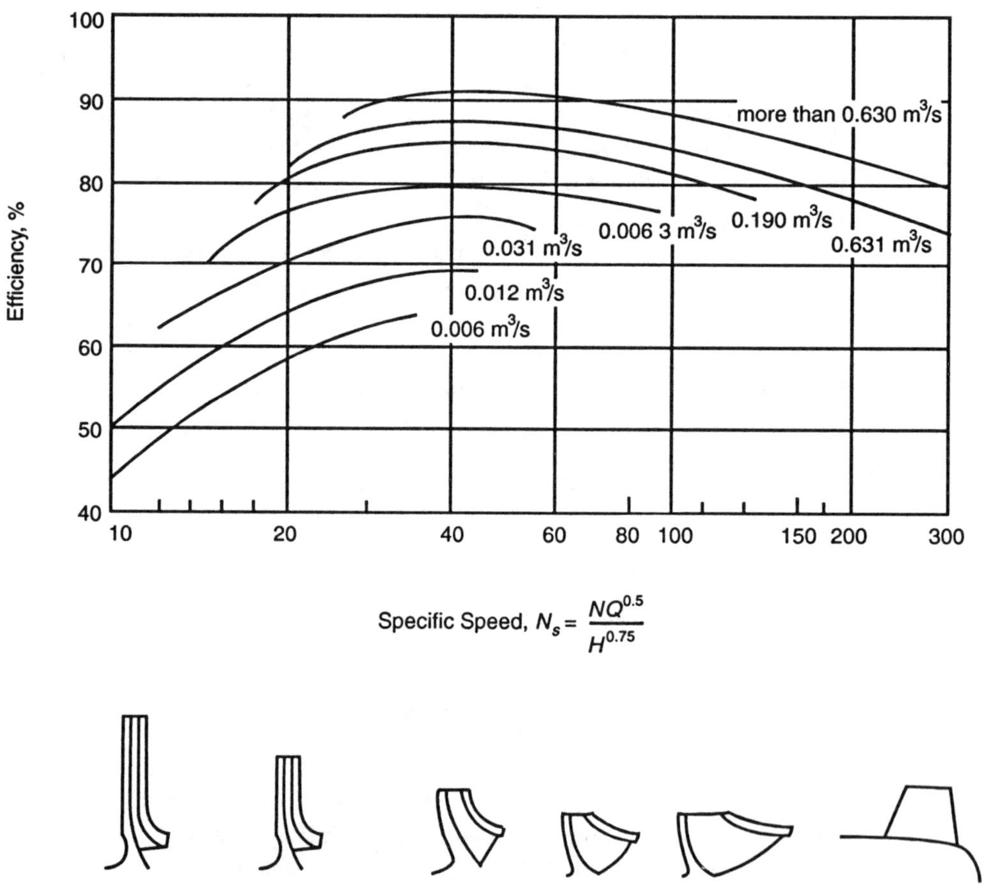

Figure 4.3 Specific speed, impeller shape, and efficiency—metric units.

units) are available with diffuser-type casings for discharge parallel with the pump shaft and are known as mixed-flow propeller pumps.

Figure 4.5 illustrates typical head capacity, power input, and efficiency curves for various specific-speed pumps. As pump-specific speed increases, the pump horsepower at shutoff (zero flow) varies from non-overloading (less than the pump horsepower at the BEP) for radial-flow pumps to extreme overloading (more than the pump horsepower at the BEP) for axial-flow pumps.

End-suction radial pumps and mixed-flow volute pumps are most commonly used in wastewater and stormwater service. Double-suction pumps are not typically used because rags can clog the pump by wrapping around the pump shaft at the eye of the impeller.

Axial-flow propeller and mixed-flow, wet-pit pumps are capable of discharging large quantities of flow at low heads and typically are less expensive than volute pumps. These pumps are common in stormwater systems but are

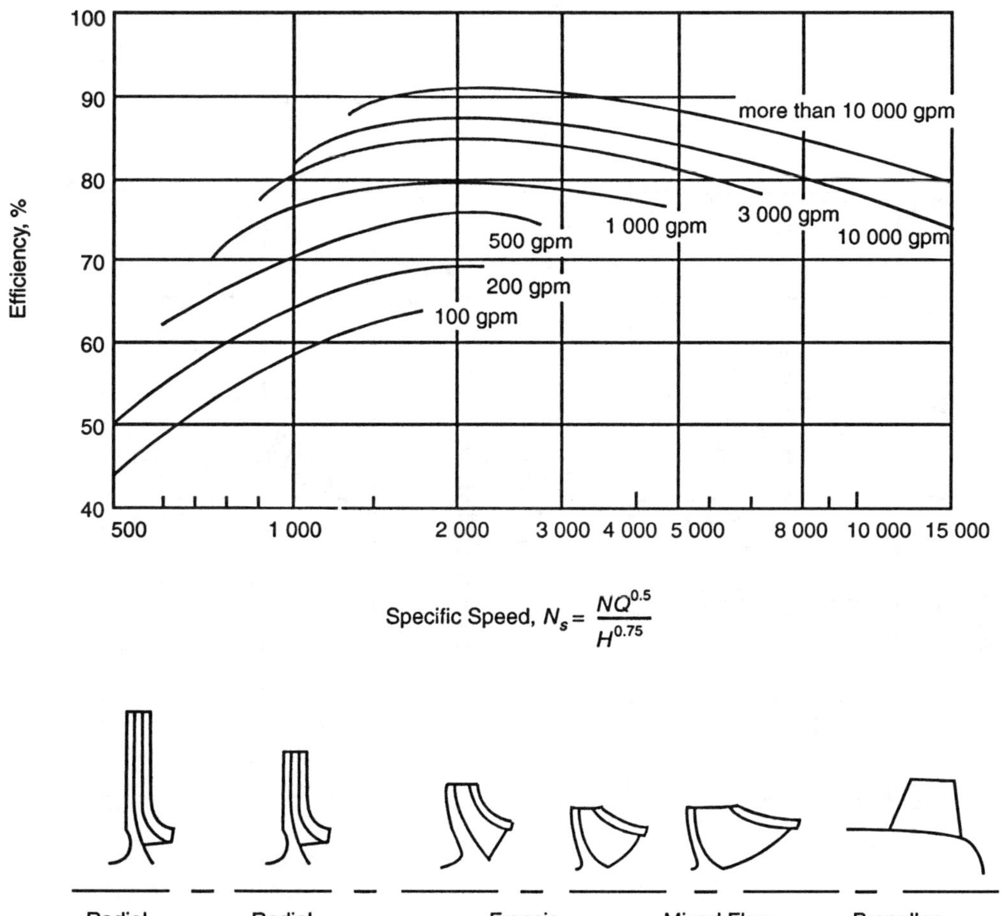

Figure 4.4 Specific speed, impeller shape, and efficiency—English units.

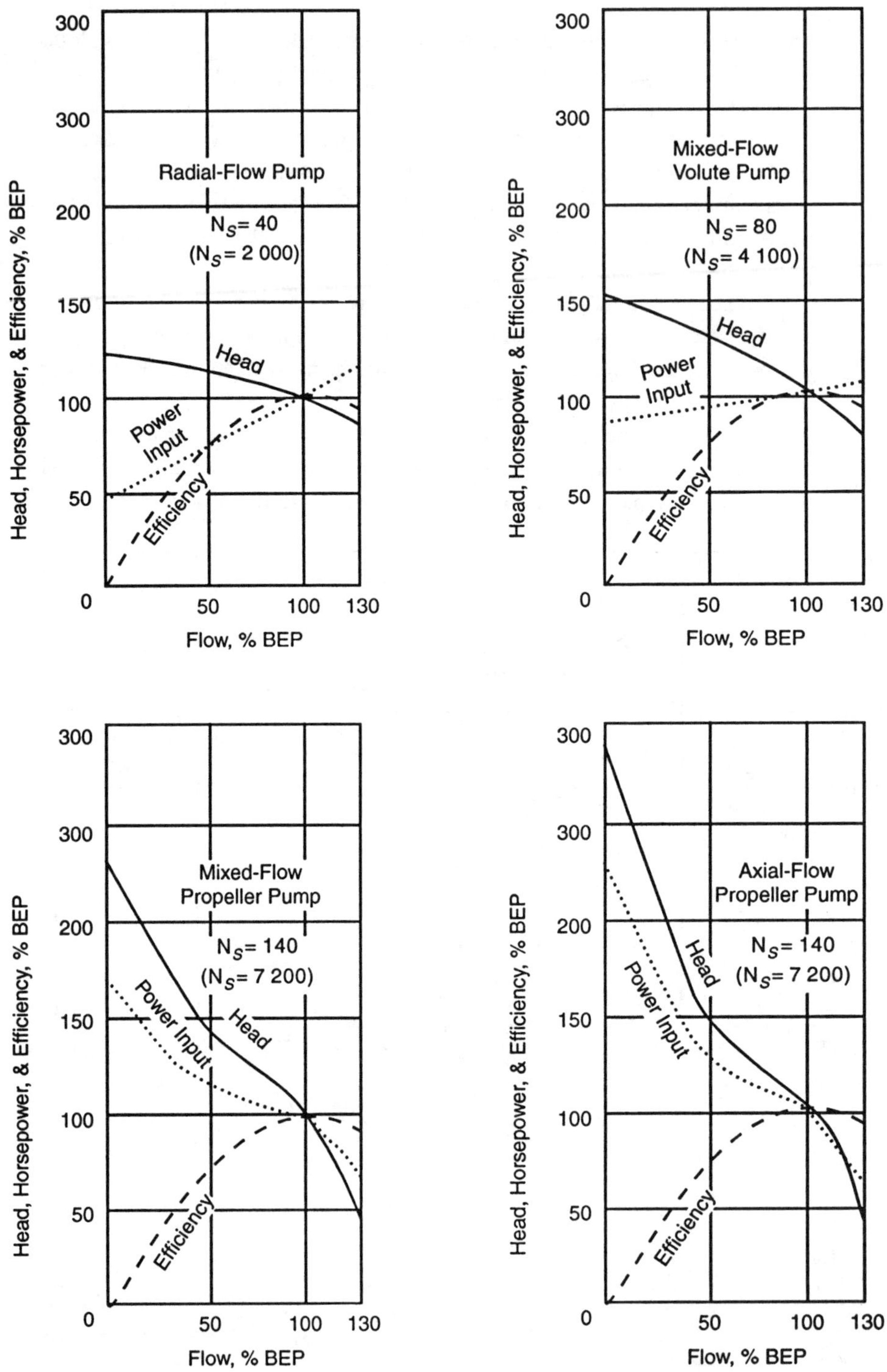

Figure 4.5 Centrifugal pump characteristic curves for various specific speeds (BEP = best efficiency point) (hp × 745.7 = W).

not typically used in wastewater systems unless well protected by screens because rags and stringy material can clog the pump by building up on the inlet or diffuser vanes.

IMPELLERS. The pump impeller is driven by the pump shaft and transfers energy to the fluid being pumped. Impellers with different specific speeds are shown in Figure 4.6.

Wastewater and stormwater pumps must be able to pass solids that enter the system. Domestic water closets can pass a 65-mm (2.5-in.) spherical solid. Therefore, many codes require wastewater pumps to pass a 75-mm (3-in.) solid. Also, rags and other stringy material enter wastewater, often presenting more of a clogging problem than solids.

Impellers are available in open, semiopen, and enclosed constructions. A semiopen impeller has one shroud on the back of the impeller and the exposed blades, or vanes, of the impeller fit closely to the stationary pump casing. This close clearance prevents recirculation from the discharge to the suction of the pump. A closed impeller has shrouds on both sides of the impeller, so close clearance is required only at the eye of the impeller between the lower shroud and the pump suction cover. An open impeller has no shrouds on the vanes.

Wastewater and stormwater radial-flow pumps typically have closed impellers to protect the edges from being eroded by grit and other abrasives. Such erosion would open up the clearance between the vane and the casing, reducing pump efficiency.

An exception is the torque-flow or vortex-type pump shown in Figure 4.7, on which the impeller is semiopen and located out of the stream of flow passing through the pump. The rotation of the impeller produces a vortex flow in the pump casing that induces flow through the pump. Because solids and stringy material do not pass through the impeller, the pump is unlikely to clog. However, these pumps are less efficient than enclosed impeller pumps.

PUMP CHARACTERISTIC CURVES. Pump characteristic curves are furnished by manufacturers to provide information on the performance of their pumps. Pumps of similar size and specific speed are produced by many manufacturers, but there are differences in pump performance because of design modifications.

Pump characteristic curves show the pump discharge head, required horsepower, and efficiency plotted against the pump discharge rate. They also indicate the net positive suction head required by the pump ($NPSH_R$) to prevent cavitation. Cavitation and $NPSH_R$ are discussed later in this chapter.

The manufacturer typically provides different diameter impellers in one size casing that cover a range of discharge heads. Therefore, the pump characteristic curve shows a series of parallel curves for a single-size pump.

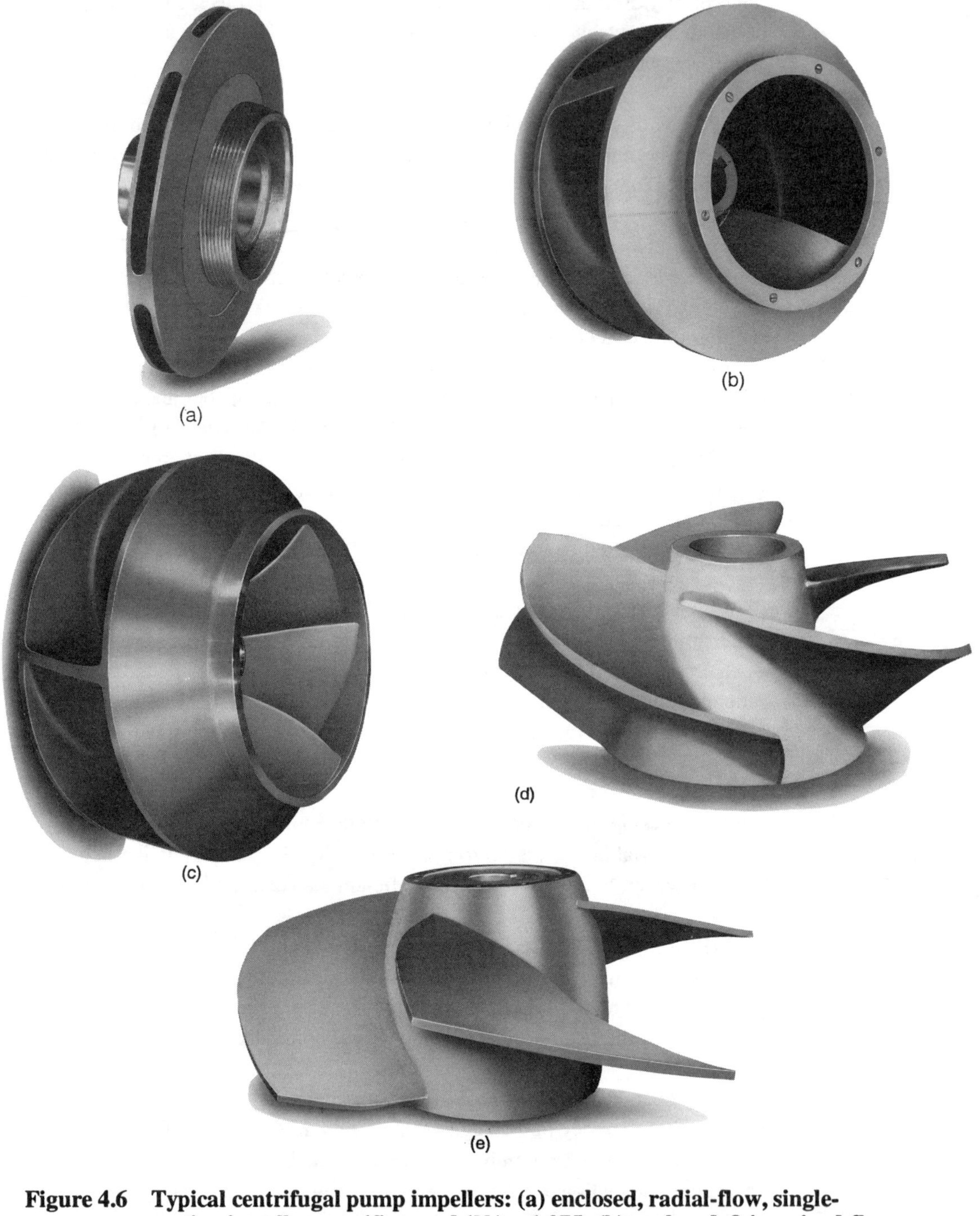

Figure 4.6 Typical centrifugal pump impellers: (a) enclosed, radial-flow, single-suction impeller, specific speed (N_s) = 1 275; (b) enclosed, 8-in. mixed-flow, volute, single-suction impeller, N_s = 3 000; (c) enclosed, 36-in. mixed-flow, volute, single-suction impeller, N_s = 3 200; (d) open, axial, mixed-flow propeller pump impeller, N_s = 7 500; and (e) open, axial-flow (propeller) impeller, N_s = 13 000. Courtesy of Ingersoll-Dresser Pump Co.

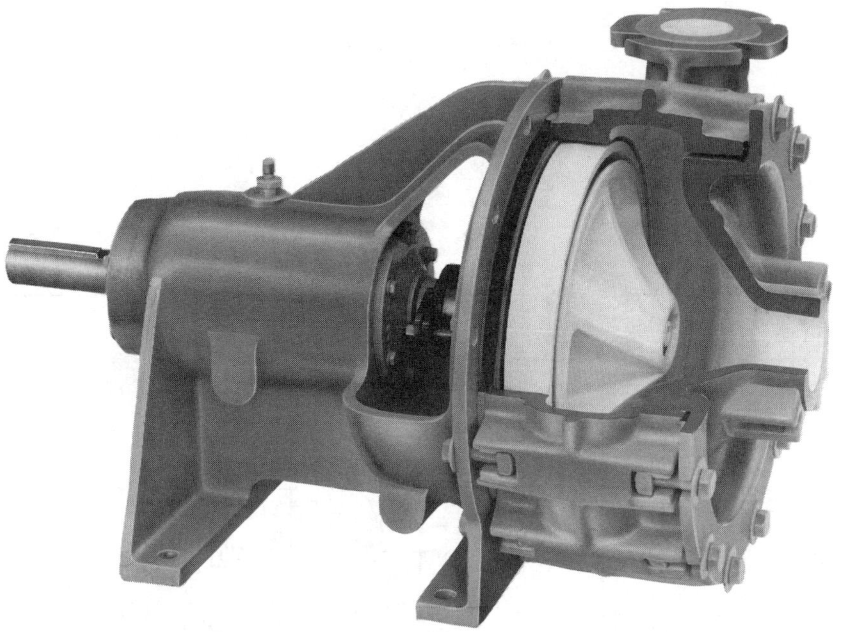

Figure 4.7 Torque flow pump with recessed impeller. Courtesy of Envirotech Specialty Pumps.

Note that the $NPSH_R$ by the largest impeller at a given flow rate is different from that by the smallest impeller at the same flow rate, and many manufacturers do not show the effect of impeller diameter on the $NPSH_R$ by the pump.

Manufacturers present characteristics in different ways. Examples of typical pump curves are shown in Figures 4.8 and 4.9. Figure 4.8 shows the $NPSH_R$, brake horsepower (BHP), and efficiency all plotted on the head-capacity curve. Figure 4.9 provides two separate curves for $NPSH_R$, one each for the largest and smallest diameter impellers. Note that the specific speed of the 600-mm (24-in.) impeller in Figure 4.8 is 38.5 (2 000), while the specific speed for the 480-mm (19-in.) impeller is 50 (2 595).

CAVITATION AND NET POSITIVE SUCTION HEAD. Cavitation in a pump occurs when the absolute pressure at the eye of the impeller is less than the vapor pressure of the fluid being pumped, causing the fluid to boil and form vapor bubbles. As these bubbles move into a zone of higher pressure, they collapse and the surrounding fluid rushes violently to fill the void, producing a hammering action that can cause vibration and pitting of the pump impeller. Cavitation results in the loss of pump capacity and efficiency and causes physical damage to the pump. When dealing with cavitation, the engineer must be familiar with net positive suction head (NPSH), which is based on absolute pressure. Two different NPSH terms are used. The first is the net

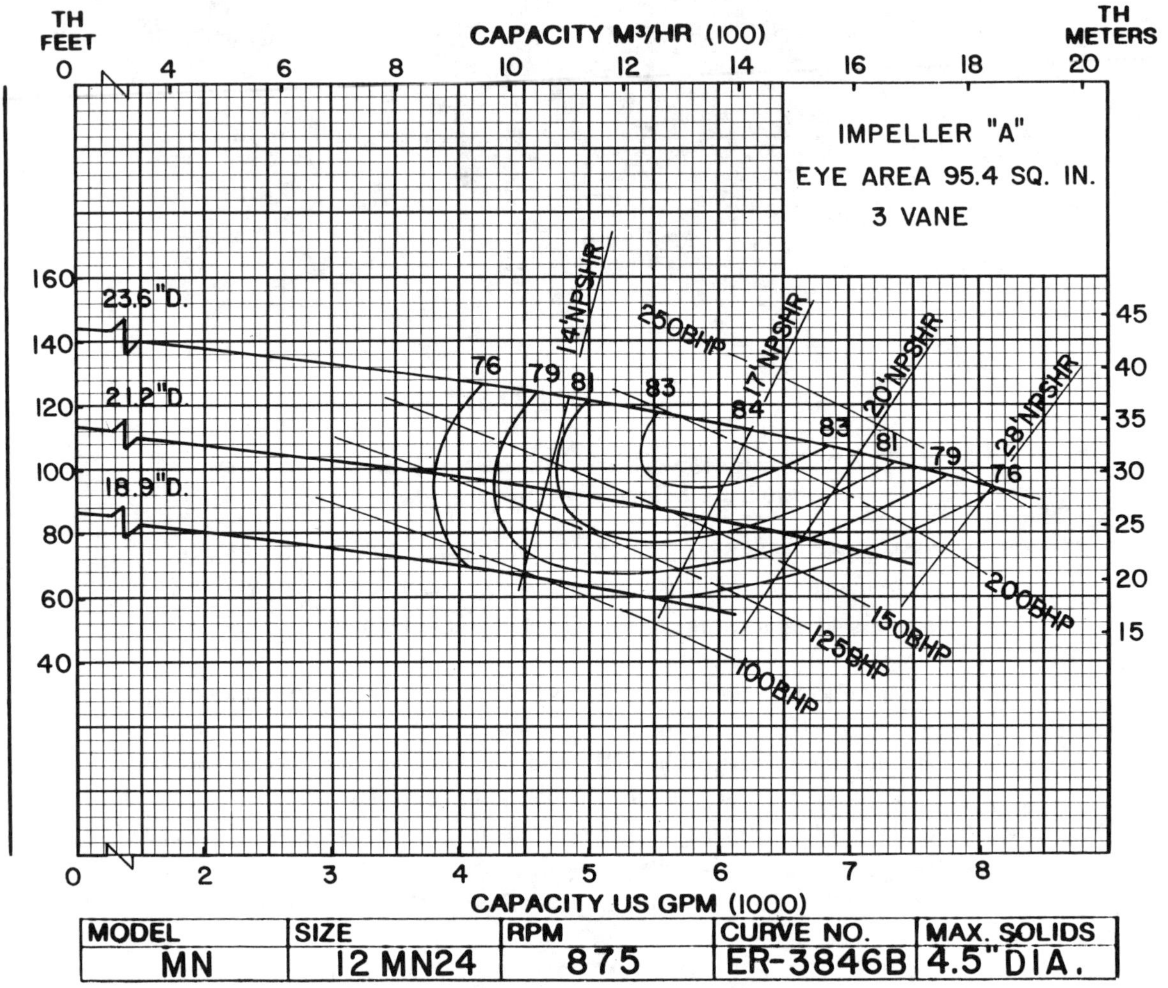

MODEL	SIZE	RPM	CURVE NO.	MAX. SOLIDS
MN	12 MN24	875	ER-3846B	4.5"DIA.

Figure 4.8 Centrifugal pump curve (superimposed $NPSH_R$ curves).

positive suction head available in the system ($NPSH_A$), which is the absolute liquid pressure produced by the system at the eye of the impeller at a given flow. The second term is the net positive suction head required by the pump ($NPSH_R$) to prevent cavitation at a given flow. Cavitation occurs when the NPSH required by the pump at the eye of the impeller is greater than the NPSH available in the system.

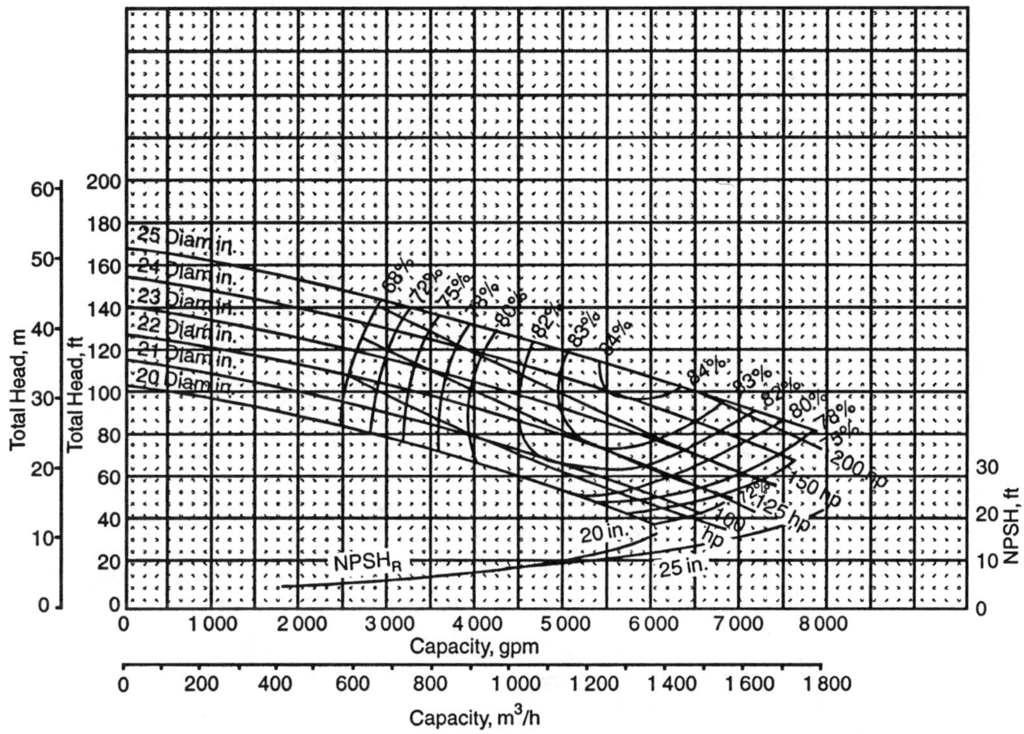

Figure 4.9 Centrifugal pump curve (separate NPSH$_R$ curves) (hp × 745.7 = W, ft × 0.304 8 = m, and in. × 25.4 = mm).

The NPSH$_A$ is the total absolute head above the vapor pressure of the fluid at the eye of the impeller. The derivation of NPSH$_A$ is illustrated in Figure 4.10 and is calculated by the following formula:

$$\text{NPSH}_A = H_A + h_{ss} - \Sigma h_{fs} - H_{vp} \tag{9}$$

Where

NPSH$_A$	=	net positive suction head available, m (ft);
H_A	=	atmospheric pressure, m (ft) of water;
h_{ss}	=	suction static head, m (ft);
Σh_{fs}	=	the sum of suction friction and minor losses, m (ft); and
H_{vp}	=	liquid vapor pressure, m (ft) of water.

Atmospheric pressure varies with elevation (Table 4.1) and liquid vapor pressure varies with the temperature of the pumped fluid (Table 4.2).

The NPSH$_R$ is a function of the impeller design and is provided by the manufacturer and typically is given on the pump curves. The NPSH$_R$ by the largest diameter impeller on the pump curve is different from the NPSH$_R$ by the smallest diameter impeller at a given flow (Figures 4.8 and 4.9).

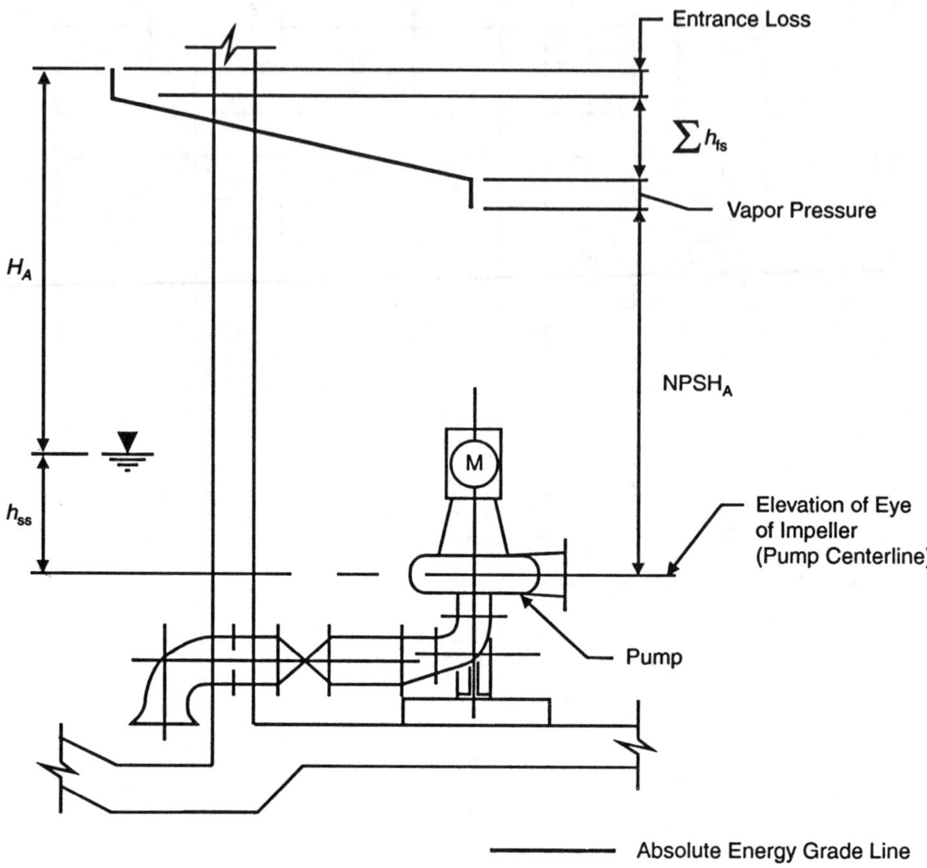

Figure 4.10 Definition sketch for system NPSH$_A$: NPSH$_A$ = net positive suction head available; H_A = atmospheric pressure; h_{ss} = suction static head; and Σh_{fs} = total suction friction losses.

The requirement to select a pump that will operate with less NPSH$_R$ by the pump than the NPSH$_A$ provided by the system is critical. The NPSH$_R$ should be checked at the extreme of pump operation with minimum static head (high wet well to low discharge water surface), with friction losses computed for new pipe.

It is recommended that 0.6 m (2 ft) be added to the NPSH$_R$ listed for the pump at a given flow. The manufacturer's NPSH$_R$ testing is done under ideal conditions, and field installation of a pump may include elbows or valves that affect the flow to the eye of the impeller. Also, there are variations in barometric pressure.

PUMP OPERATING RANGE. The operating range of a pump is limited by hydraulic and mechanical considerations.

Table 4.1 Atmospheric absolute pressure at various altitudes.

Altitude above sea level, m	Atmospheric pressure, m of water at 20°C	Altitude above sea level, ft	Atmospheric pressure, ft of water at 70°C
0	10.3	0	34.0
300	10.0	1 000	32.8
600	9.6	2 000	31.6
900	9.3	3 000	30.5
1 200	9.0	4 000	29.4
1 500	8.6	5 000	28.3
1 800	8.3	6 000	27.3
2 100	8.0	7 000	26.2
2 400	7.7	8 000	25.2
2 700	7.4	9 000	24.3
3 000	7.1	10 000	23.4

Table 4.2 Vapor pressure at various temperatures.

Fluid temp., °C	Vapor pressure, m of water	Fluid temp., °F	Vapor pressure, ft of water
1.7	0.06	35	0.2
4.5	0.09	40	0.3
10	0.12	50	0.4
15	0.18	60	0.6
21	0.27	70	0.9
27	0.37	80	1.2

Like most mechanical equipment, a pump operates best at its BEP, where flow is streamlined, reducing turbulence, erosion of parts, and hydraulic loads on bearings. Operating at their BEP, pumps can function many years with minimum maintenance. Unfortunately, most wastewater pumps must operate over a range of pressures and flows at points considerably removed from the BEP.

Cavitation is a potential problem when a pump operates on its head-capacity curve well beyond its BEP. At this point, the pump $NPSH_R$ increases rapidly as pump discharge increases.

If the pump operates near shutoff head (zero flow), recirculation within the pump becomes a problem. At low flow, flow through the impeller is not smooth so that as the flow passes through the impeller, increasing the head, a portion of the flow reverses to return to the lower pressure at the eye of the impeller. This recirculation causes vibration in the pump and results in cavitation-like damage to the impeller. The flow at which this recirculation starts is closer to the BEP for high, specific-speed pumps than for low, specific-speed pumps.

Mechanically, pump thrust and radial loads increase as the flow moves away from the BEP. These increased forces reduce bearing life and increase shaft deflection that can, in turn, result in wear ring contact and failure of mechanical seals or packing and even the shaft.

For these reasons, it is good practice to limit the range of operation on the pump curve to between approximately 60 and 125% of the best efficiency flow. This range can be extended on low, specific-speed pumps and for lower speeds and heads, but care should be taken when operating outside this range. For systems that require pump operation over a wide range of flows, the pump manufacturer should be consulted for the recommended range of operation.

AFFINITY LAWS. Affinity laws are used to predict centrifugal pump performance at different speeds and with different diameter impellers. These laws predict the slope of the head-capacity curve, the power required by a pump with great accuracy for changes in speed, and changes in impeller diameter (with less degree of accuracy).

For changes in speed, but with the impeller diameter remaining constant, the affinity law becomes

$$\frac{Q_2}{Q_1} = \frac{N_2}{N_1} \text{ or } Q_2 = Q_1 \frac{N_2}{N_1} \tag{10}$$

$$\frac{H_2}{H_1} = \frac{(N_2)^2}{(N_1)^2} \text{ or } H_2 = H_1 \frac{(N_2)^2}{(N_1)^2} \tag{11}$$

$$\frac{P_2}{P_1} = \frac{(N_2)^3}{(N_1)^3} \text{ or } P_2 = P_1 \frac{(N_2)^3}{(N_1)^3} \tag{12}$$

Where

Q_1 = pump flow rate, L/s (gpm) at speed 1;
Q_2 = pump flow rate, L/s (gpm) at speed 2;
N_1 = pump rotative speed, r/min (rpm);
N_2 = pump rotative speed, r/min (rpm);
H_1 = pump head, m (ft) at speed 1;
H_2 = pump head, m (ft) at speed 2;
P_1 = pump power required, kW (hp) at speed 1; and
P_2 = pump power required, kW (hp) at speed 2.

When using affinity laws, computed flow, head, and power are called corresponding points. To develop a head-capacity curve for a different speed, enough corresponding points should be calculated to plot the curve at the new speed. The efficiency of corresponding points remains essentially the same for changes in pump speed.

Figure 4.11 indicates the use of affinity laws to develop pump curves for different speeds for the pump in Figure 4.8. The pump operating range and operating envelope are also shown.

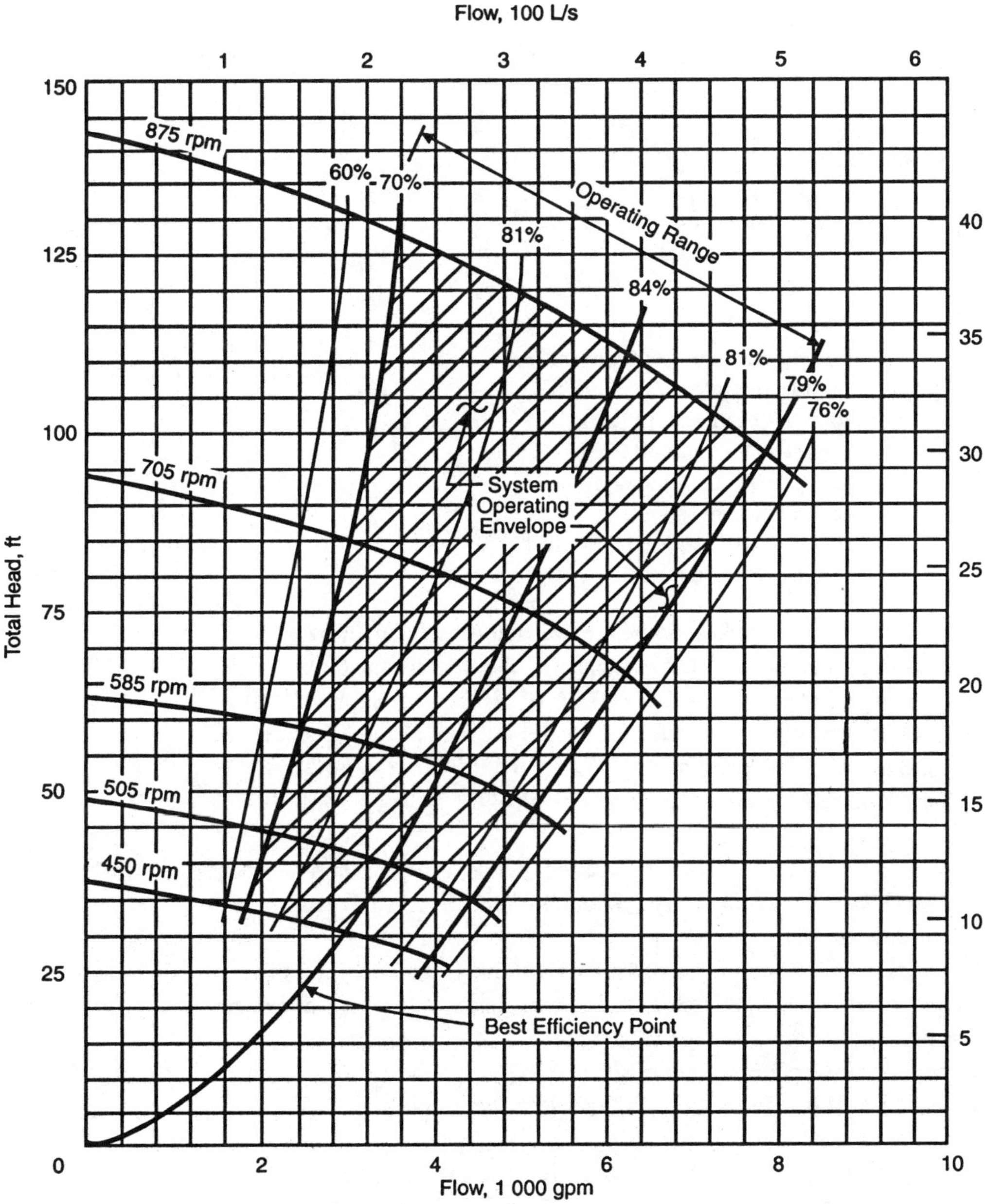

Figure 4.11 Pump operating range and operating envelope.

Corresponding points also can be developed for a change in impeller diameter when the speed is held constant. However, results are less accurate and the efficiency among corresponding points will vary. For changes in impeller diameter, with the speed held constant, the affinity law becomes

$$\frac{Q_2}{Q_1} = \frac{D_2}{D_1} \text{ or } Q_2 = Q_1 \frac{D_2}{D_1} \tag{13}$$

$$\frac{H_2}{H_1} = \frac{(D_2)^2}{(D_1)^2} \text{ or } H_2 = H_1 \frac{(D_2)^2}{(D_1)^2} \tag{14}$$

$$\frac{P_2}{P_1} = \frac{(D_2)^3}{(D_1)^3} \text{ or } P_2 = P_1 \frac{(D_2)^3}{(D_1)^3} \tag{15}$$

Where

Q_1	=	pump flow rate, L/s (gpm) at D_1;
Q_2	=	pump flow rate, L/s (gpm) at D_2;
D_1	=	impeller diameter, m (in.);
D_2	=	impeller diameter, m (in.);
H_1	=	pump head, m (ft) at D_1;
H_2	=	pump head, m (ft) at D_2;
P_1	=	pump power required, kW (hp) at D_1; and
P_2	=	pump power required, kW (hp) at D_2.

The pump curve indicated in Figure 4.8 illustrates the accuracy of using affinity laws on changes in diameter. For instance, the BEP for the 600-mm (24-in.) diameter impeller in Figure 4.8 is 390 L/s (6 200 gpm) at 35 m (110 ft) with 84% efficiency. The calculated BEP for the 480-mm (19-in.) diameter impeller is 315 L/s (5 000 gpm) at 22 m (72 ft), but the curve shows the BEP at 300 L/s (4 600 gpm) at 20 m (65 ft) with only 78% efficiency.

PUMP CONSTRUCTION. The construction of a centrifugal pump depends on the pump location (wet pit or dry pit) and, to some degree, on the size, capacity, and duty of the pump.

Pumps are installed with either vertical or horizontal shaft configuration. Wet-pit pumps almost always are vertical. The majority of dry-pit stormwater and wastewater pumps are arranged for vertical mounting, which requires less space than horizontal installation and results in the drive being located above the pump. This reduces the possibility of flooding the drive in the event of a catastrophe.

Wet-pit pumps include mixed-flow propeller and axial-flow column pumps and submersible pumps. Axial-flow column pumps typically are driven by drives located above the wet well, with shafts that pass through the discharge head.

Submersible pumps can be both axial-flow pumps or radial-flow pumps that are driven by submersible motors. These pumps are becoming more common and are used in both wet-pit and dry-pit configuration. When used as wet-pit pumps, they are arranged for removal from the wet pit for maintenance, thereby eliminating the cost of a dry well. When used in a dry pit, the submersible pump eliminates the need for intermediate shafts to connect a motor located above flood level with the pump and allows the station to continue in operation in the event of catastrophic flooding.

Radial-flow Volute Pumps. The major parts of a radial-flow volute pump are the rotating assembly, the casing, and the bearing frame, or bearing housing. Pumps are classified as close coupled, direct connected, or shaft driven.

Close-coupled pumps have a rotating assembly and casing but do not have a bearing frame. The impeller is mounted directly on the shaft of the motor and the casing of the pump is connected to the motor frame. Motor bearings serve as the pump bearings. Close-coupled pumps typically are not used in municipal service. The exceptions to this are small pumps in package pumping stations, where space is limited, and submersible pumps.

Direct-connected pumps have a rotating assembly, a casing, and a bearing frame. The pump shaft is connected directly to the output shaft of the drive through a flexible coupling. The bearing frame of the pump contains the shaft bearings. Figure 4.12 shows a typical dry-pit, direct-connected wastewater pump.

The external shaft-driven, vertical, dry-pit pump is similar to the direct-connected pump except that the pump drive motor is located above and is connected to the pump by a drive shaft. A shaft-driven pump is shown in Figure 4.13.

The rotating assembly includes the pump shaft, impeller, shaft sleeves, impeller wear rings, mechanical seals, bearings, and all pump parts, whether rotating or stationary, that must be assembled over the shaft.

When servicing a pump, the rotating assembly and the bearing frame typically are removed from the pump casing and taken to a shop where servicing is more convenient. For this reason, lifting devices to remove the rotating assembly and maneuvering space around the pump should be included in station design.

Replaceable wear rings typically are installed on the impeller and the casing at the inlet to the impeller to protect them from wear and reduce recirculation from the higher discharge pressure to the lower suction pressure. Because one rotates and the other is stationary, there must be clearance between the impeller and the casing, so there will be some recirculation. Abrasives in wastewater would erode the impeller and the casing without the wear rings, which should be replaced when clearances become excessive.

Wear rings can be designed for either close axial fit or close radial fit. Smaller pumps typically are furnished with close axial-running clearance, requiring adjustment of the clearance by axial adjustment of the impeller. Larger, more expensive pumps are furnished with radial or "L" shaped rings having close radial clearance. The pump manufacturer should be consulted for recommended wear ring clearances.

The pump shaft transmits torque from the drive to the impeller and is supported by bearings mounted in the bearing frame. The shaft is subject to wear at the stuffing box, so replaceable shaft sleeves are installed on the shaft where it passes through the stuffing box to prevent damage to the shaft.

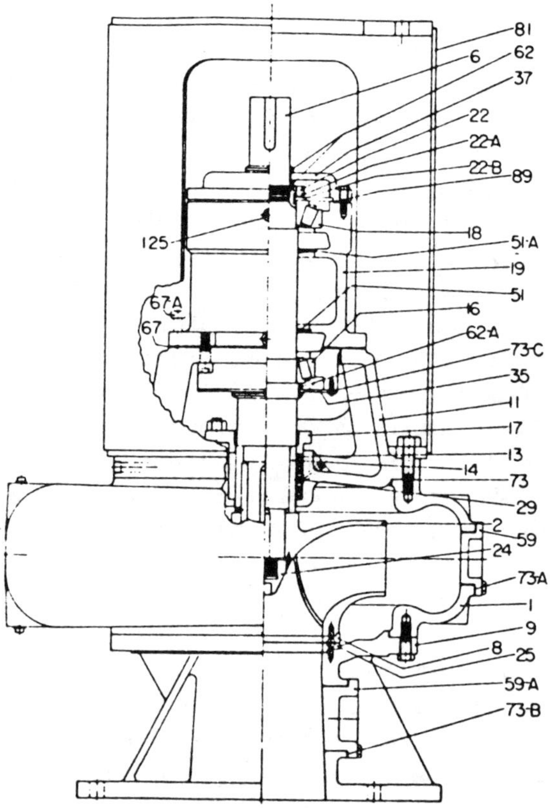

STATIONARY PARTS						ROTATING PARTS		
Ref. No.	No. of Pieces	Name of Part	Ref. No.	No. of Pieces	Name of Part	Ref. No.	No. of Pieces	Name of Part
1	1	Casing	51A	1	Grease Retainer-Outboard	2	1	Impeller
9	1	Suction Head	59	1	Handhole Cover-Casing	6	1	Shaft with Keys
11	1	Stuffing Box Head	59A	1	Handhole Cover-Suct. Hd.	8	1	Impeller Wrg. Ring
13	5	Packing Rings	67	1	Adjusting Shim Set	14	1	Shaft Sleeve
17	1	Gland	67A	2	Adjusting Bolt	16	1	Line Bearing
19	1	Bearing Frame	73	2	Ankorite Gasket	18	1	Thrust Bearing
25	1	Suction Head Wrg. Ring	73A	1	Rubber Gasket	22	1	Bearing Locknut
29	2	Seal Cage Halves	73B	1	Rubber Gasket	22A	1	Lockwasher
35	1	Line Brg. Cover	73C	1	Paper Gasket	22B	1	Bearing Washer
37	1	Thrust Brg. Cover	81	1	Motor Pedestal	24	1	Impeller Nut
51	1	Grease Retainer-Inboard	89	1	"O" Ring Seal	62	2	Seal Ring
			125	2	Alemite Fittings	62A	2	Seal Ring

Figure 4.12 Typical vertical, dry-pit volute pump.

> ⚠ **WARNING**
>
> Do not operate this machine without protective guard in place. Any operation of this machine without protective guard can result in severe bodily injury.
>
> -A- Supplied by FMPC -B- Supplied by others

Notes:

(1) All flanges 55 kg (25g ANSI) drilling unless noted.

(2) All dimensions are in inches unless noted.

(3) Dimensions reflect usable shaft length.

(4) Bases are designed to have full contact with grout or a sole plate grouted in place.

(5) Install drive shaft per manufacturer's recommendation.

(6) Not for construction, installation, or application purposes unless certified. Dimensions shown may vary due to normal manufacturing tolerances.

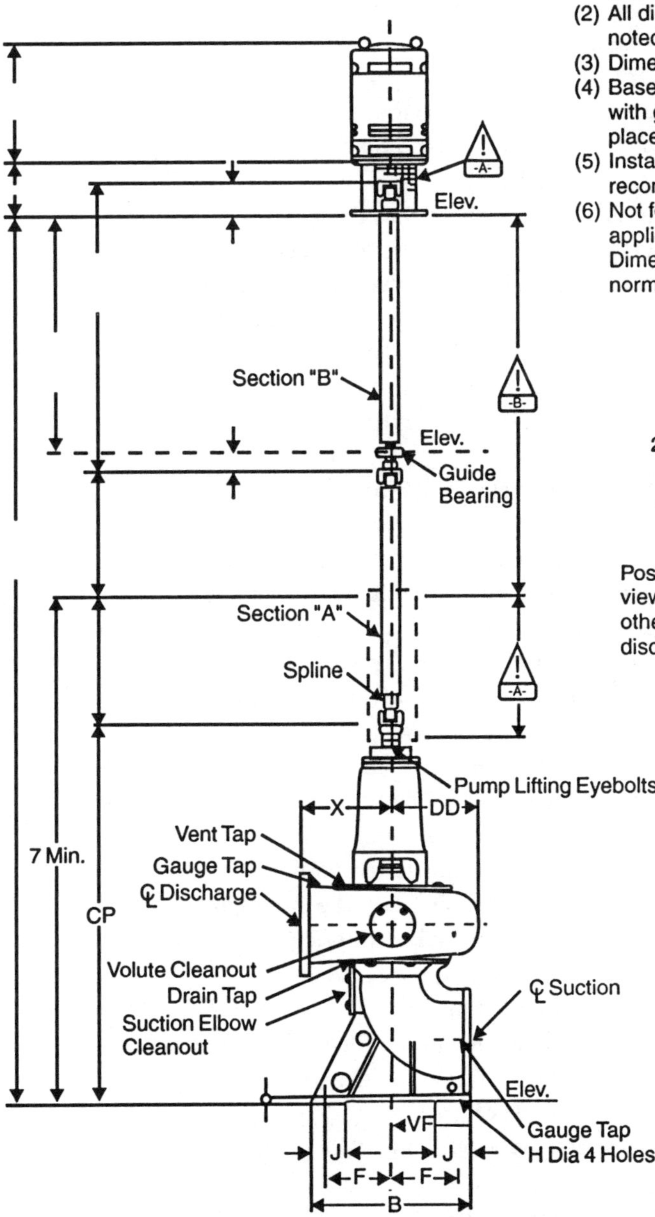

Available discharge positions

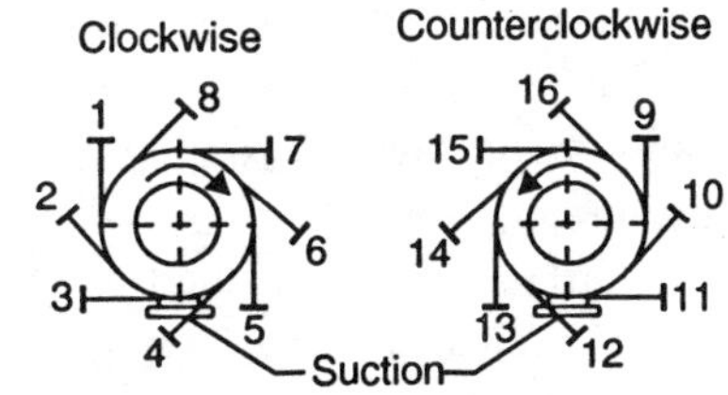

Positions #1 or #9 are standard when viewed from the driver end unless otherwise specified. Clockwise rotation discharge position #1 shown.

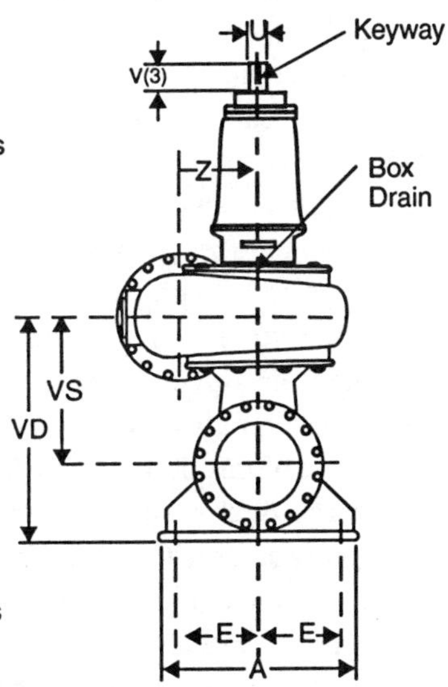

Figure 4.13 Typical vertical, shaft-driven volute pump (in. × 25.4 = mm).

Where the drive is mounted above flood level, intermediate shafting is required to connect the drive to the pump. Depending on the distance, more than one section of shafting might be needed, and intermediate shafting must be selected carefully to eliminate destructive vibration. Vibration can be lateral (translational, movement from side to side or up and down), or torsional (twisting).

All shafts have a natural frequency, or critical speed, that is dependent on length, tube dimensions, and tube material. The pump system produces lateral vibrations at several operating frequencies, known as exciting frequencies. Vibrations caused by imbalance occur at the pump operating speed; those caused by angular displacement in the shaft U-joint occur at two times the rotating speed; and vane-passing vibration occurs at the multiple of the rotative speed times the number of vanes in the impeller.

It is critical that exciting frequencies not coincide with the critical speed of the shaft. A conservative design requires that none of the exciting frequencies are higher than 75% of the critical speed or within ±10% of the half critical speed.

Torsional stress also should be investigated. For a conservative design, torsional natural frequency should not be closer than 25% of an exciting frequency.

The selection of shafting for a variable-speed pump is complicated because of the range of exciting frequencies produced over the speed range of the pump.

Until recently, most shafts were made of steel tubing. Shafts now are available in composite material of graphite and/or glass fibers in an epoxy thermoset plastic. These shafts are more expensive and difficult to handle than steel but have a higher critical speed and can be furnished in longer lengths.

The pump casing is the stationary housing around the impeller and includes the volute, suction cover, and stuffing box cover. The volute casing is spiral shaped, collects the discharge from the impeller, and directs the flow to the discharge pipe. The suction cover and stuffing box cover are attached to the volute to enclose the impeller. One point in the volute, known as the cut water, is close to the outside diameter of the impeller. The flow area of the volute increases from the cut water to the discharge connection.

Hydraulic forces produce a radial thrust on the impeller proportional to the pump head and the projected area of the impeller. This thrust is a minimum at BEP, but increases rapidly as the flow increases beyond or drops below the BEP, and is maximum at shutoff head. Hydraulic Institute Standards (1975) provide data for approximating the radial loads on pumps.

Because of this thrust, a few large, high-head pumps are fitted with double volutes. In a double-volute pump, the volute is furnished with two cut waters 180 degrees apart with a vane from the second cut water to the discharge. This arrangement divides the flow between the two sides and the radial thrusts tend to cancel each other.

The suction cover connects the suction pipe to the pump casing. On vertical pumps up to approximately 400 mm (16 in.), the suction cover can include a suction elbow and serve as a support base.

Handholes are required on the suction pipe and volute to allow inspection of the pump and removal of solids lodged between the impeller vanes, clogging the pump. Handholes should be as large as possible and the interior of the cover must match the contour of the pump.

The stuffing box cover on volute casings forms the back of the casing and contains a stuffing box to prevent leakage where the pump shaft passes through the cover.

The stuffing box can be sealed with packing or a mechanical seal. One or two rows of packing are located in the stuffing box close to the impeller, with an adjacent lantern ring. Three or more rows of packing are located between the lantern ring and the outside of the stuffing box. The packing is held in the stuffing box by an adjustable gland that maintains pressure on the packing.

Because wastewater contains abrasives and packing requires lubrication, the packing must have an outside source of lubricant at a higher pressure [69 kPa (10 psi) or greater] than that in the pump. This ensures that leakage will be from the lubricant to the pump to keep abrasives out of the packing. This lubricant enters the stuffing box at the lantern ring, with a small amount of lubricant flowing into the pump casing and a small amount passing along the shaft to the atmosphere.

The most common lubricants are water or grease. If water is used, it must be from a clean source (if potable water, the source must be protected by an air break or be supplied through a backflow preventer, though some local codes do not allow backflow preventers). A typical packed stuffing box with water lubrication is shown in Figure 4.14.

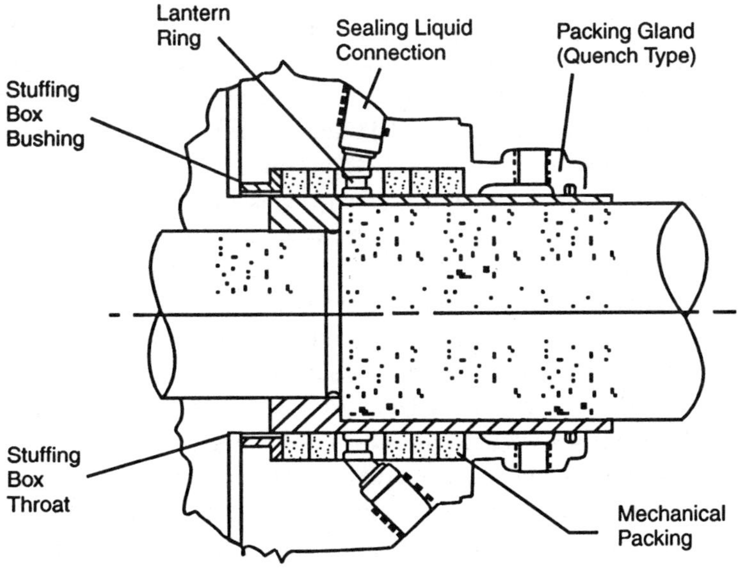

Figure 4.14 Stuffing box with packing.

Spring-loaded or weight-loaded lubricators are available if grease lubrication is used. These lubricators should be controlled to provide lubricant only when the pump is operating.

If a packed stuffing box is used, the stuffing box will have a continuous, slight leakage, which should be piped to a drain.

Mechanical seals also are used to seal the stuffing box. A mechanical seal has a lapped rotating surface and a lapped stationary surface, and the smooth surfaces practically eliminate leakage. The faces are lubricated by a small amount of fluid that passes between the surfaces.

Mechanical seals are available in internal and external configuration and single and double seal arrangements. Because of abrasives in the wastewater, double mechanical seals are recommended. A typical double mechanical seal is shown in Figure 4.15.

Double mechanical seals consist of two seals arranged back to back, with one seal facing the impeller and the second facing the exterior of the pump. The space between the two seals must be provided with seal water to lubricate and cool the surface between sealing surfaces. Again, this lubricant must be at a higher pressure than the pumped wastewater. A small amount of this lubricant must flow through the seal to provide cooling, and this must be piped to a drain.

There are advantages and disadvantages to mechanical seals. Mechanical seal failures are complete and result in considerable leakage until the pump is shut down. The packed stuffing box leakage gradually increases as the packing wears.

The replacement of a conventional mechanical seal requires the disassembly of the pump to place the mechanical seal over the shaft of the pump. Packing can be replaced while the pump is assembled. Split mechanical seals that allow seal replacement without pump disassembly are a recent development but are expensive and require more space. The pump manufacturer should be consulted to ensure that the pump can be fitted with split mechanical seals.

The bearing frame on direct-connected and shaft-driven pumps provides support for the pump shaft and contains the pump bearings.

Hydraulic forces impose both radial- and axial-thrust loads on the impeller. The discharge pressure acts on both sides of an enclosed impeller. These loads cancel each other, except at the eye of the impeller, where one side of the impeller is acted on by the suction pressure while the other side sees the discharge pressure. This imbalance produces thrust.

In large, high-head pumps, the axial thrust can be reduced by installing pump-out vanes on the back of the impeller or installing a second set of wear rings on the back of the impeller, opposite the edge of its eye, and drilling a balancing port from the inlet to the chamber inside the back wear rings. This allows the suction pressure to act on both sides of the impeller, reducing axial thrust.

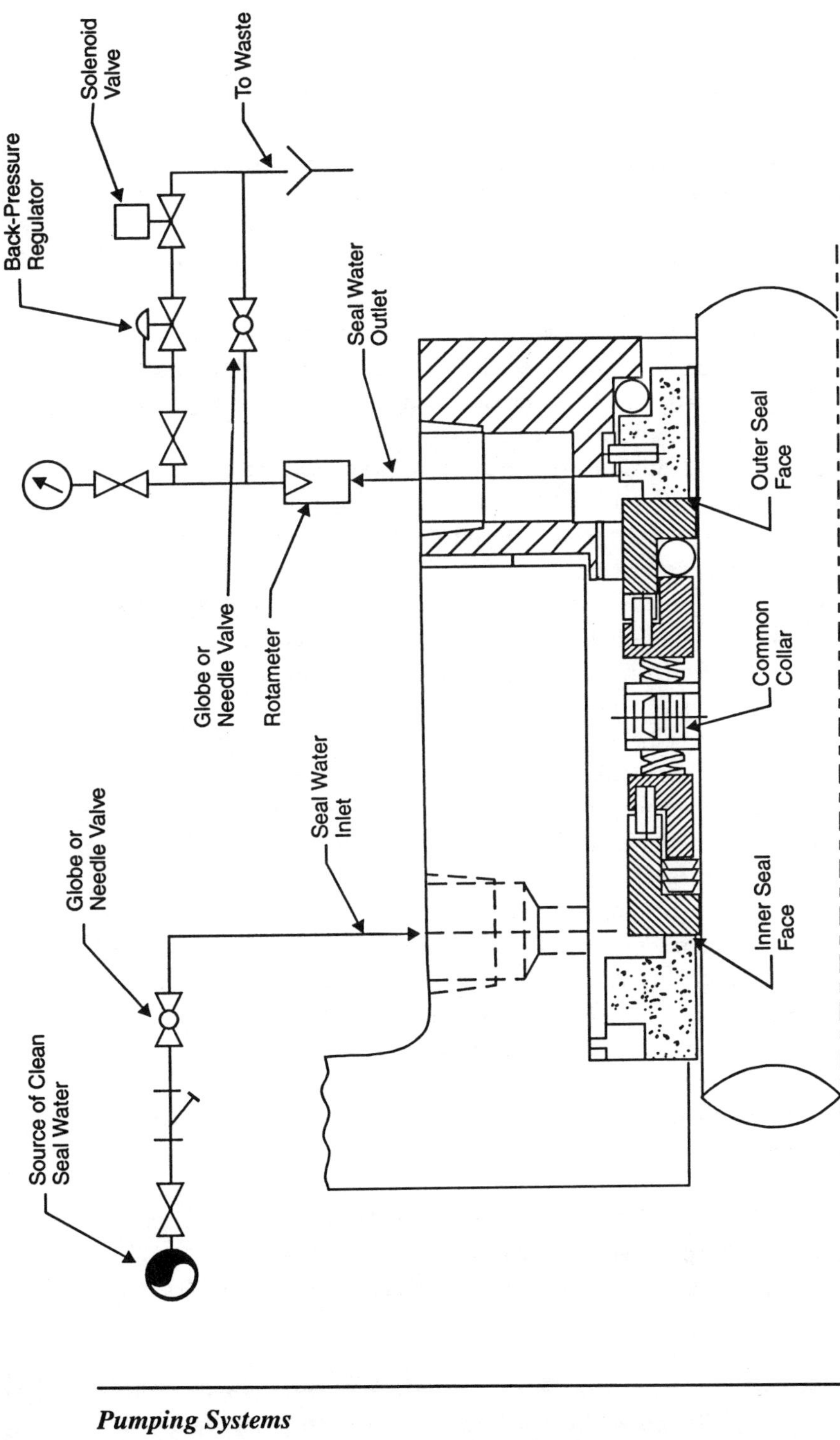

Figure 4.15 Typical double mechanical seal.

The outboard bearing, the one farthest from the impeller, is designed for both axial- and radial-thrust loads. The inboard bearing is designed for radial loads.

Most new volute pumps have anti-friction-type ball or roller bearings, with the roller bearings only used on larger pumps with heavier loads. These antifriction bearings can be lubricated with oil or, more commonly, grease.

The thrust load of some large [1-m (35-in.) and larger] high-head pumps can exceed the load capacity of antifriction bearings. For these applications, a Kingsbury-type thrust bearing can be mounted in the pump or motor.

Axial-flow Column Pumps. Axial-flow, wet-pit column pumps differ from radial-flow pumps in many ways. The major parts of an axial-flow, wet-pit pump include the rotating assembly, suction bell, discharge bowl or bowls, column, and discharge head and drive. An axial-flow, wet-pit pump is shown in Figure 3.8.

The impellers are high, specific-speed, mixed-flow, or propeller-type impellers. Mixed-flow impellers can be semiopen or enclosed and discharge flows at an angle from the shaft. The propeller impeller is an open-type impeller and discharges flow parallel to the shaft.

These pumps can be single stage or multistage as necessitated by the system discharge head.

The suction bell directs the flow to the inlet of the impeller and houses the shaft bottom bearing of the pump.

The discharge bowl encloses the impeller and contains the pump bearings. Discharge bowls for mixed-flow impellers are provided with diffuser vanes that collect the flow from the impeller and guide it axially up the column or to the inlet of the next stage impeller.

The column connects the discharge bowls to the pump discharge head, which can be above or below the floor that supports the pump in the wet well. The discharge head diverts the flow from parallel to the pump shaft to perpendicular to the shaft for connection to external piping.

The pump shaft passes through a stuffing box in the discharge head and also passes through the column pipe.

Axial-flow, wet-pit pumps typically are furnished with sleeve bearings designed for radial loads only. Axial thrust is absorbed by a thrust bearing in the drive and, because the pumped fluid contains grit, the shaft and bearings are mounted in a shaft-enclosing tube. To lubricate the sleeve bearings, clean water or oil is admitted through the top of this tube or grease is supplied to the bearings through grease tubes. The enclosing tube is structurally stabilized in the column pipe to maintain alignment.

Construction Materials. Considerations in selecting construction materials include the fluid being pumped, abrasive material in the fluid, and cavitation potential. Raw wastewater and stormwater are neutral fluids (at a pH of 7) at temperatures between 5 and 30°C (40 and 85°F). However, wastewater and stormwater do contain abrasive materials (grit) that cause wear in pumps.

The casing and bearing frame must be of heavy construction to absorb the force of the fluid being pumped and the loads imposed on the bearings.

Almost without exception, the casing and bearing frame are made of cast iron. Higher-strength ductile iron sometimes is used for larger, high-head pumps. Cast steel occasionally is used for the casing in high-head pumps on flows containing large quantities of grit to reduce erosion in the casing.

The shaft furnished in most pumps is of carbon steel, which has the strength needed to deliver the input power to the impeller.

Sleeves are installed on the shaft where it passes through the stuffing box, protecting the shaft from abrasive wear. Shaft sleeves should be hardened to 350 to 450 Brinell Hardness (BNH), and those on packed pumps are typically of 400 series stainless steel or its equivalent casting alloy. This material is resistant to wear and is machinable. Shaft sleeves of pumps with mechanical seals are not exposed to the abrasive wear of a packed pump and can be of 300 series stainless steel.

Impeller material considerations include the quantity of grit in the wastewater being pumped, the flow rate in relation to best efficiency, and the cost of the material.

Most wastewater service impellers are made of cast iron, which is the least expensive material and provides adequate service where the grit load is not excessive.

For large, higher-head pumps, the extra cost of materials more resistant to erosion and cavitation damage is justifiable. Table 4.3 contains information from *Pump Handbook* (Karassik *et al.*, 1976), which lists materials in order of increasing resistance to cavitation and abrasive wear damage. In each case, cast iron displays the least resistance. The extra cost of higher-resistant materials must be weighed against the benefits of longer impeller life.

Wear rings are installed to reduce recirculation in a pump and are subject to erosion and galling.

Table 4.3 Material resistance to abrasion and cavitation damage.

	Cavitation resistance[a]		Abrasion resistance[a]
1	Cast iron	1	Cast iron
2	Bronze	2	Bronze
3	Cast steel	3	Manganese bronze
4	Manganese bronze	4	Nickel/aluminum/bronze
5	Monel	5	Cast steel
6	400 series stainless steel	6	300 series stainless steel
7	300 series stainless steel	7	400 series stainless steel
8	Nickel/aluminum/bronze		

[a] Listed in order of increasing resistance (cast iron least resistant).

Stainless steel of the 400 series or its equivalent casting alloy typically is used for the aforementioned parts. The impeller wear ring should be hardened to 250 to 350 BNH and the casing wear ring should be hardened to approximately 100 BNH harder than the impeller ring to reduce the potential for galling.

POSITIVE DISPLACEMENT PUMPS

Most positive displacement pumps have limited application in stormwater and wastewater pumping because of close clearances and wear on the pump parts. The two types of positive displacement pumps commonly used are the Archimedes screw pump and the pneumatic ejector, which are at the opposite ends of flow rates. The screw pump typically is used for large flows and the pneumatic ejector is typically used for small flows.

Two other significant types of pumps are the airlift pump and the progressive cavity pump.

ARCHIMEDES SCREW PUMPS. One of the oldest types of pumps, this is based on the Archimedes screw principle in which a revolving spiral, rotating in an inclined trough, pushes the flow up the trough.

Screw pumps are lift pumps that draw water from an open channel, lift it, and discharge to another open channel at a higher elevation. The screw pump has two distinct advantages when pumping stormwater or wastewater: (1) it can pump large solids and stringy material without clogging and (2) it pumps variable flow rates over a wide range of flows without complicated control equipment, thereby matching incoming flow.

Pumps are driven by constant-speed motors through a gear box at speeds normally between 25 and 75 r/min (rpm), and screw pump drive motors by design do not overload. The capacity of the screw pump increases as the flow level entering the screw rises until it reaches the fill point, which is the approximate center of the screw. Though pump capacity cannot increase as flow level rises above the fill point, the pump head decreases, reducing pump horsepower. If the drive motor is sized for the pump capacity at the fill point, the motor cannot be overloaded.

Screw pumps are either open or enclosed. Open screw pumps have screw flights welded to the outside of a central torque tube. An open screw pump installation is shown in Figure 4.16.

Open screw pumps are available with diameters from 0.6 to 3.6 m (24 to 144 in.) and capacity ranges from 60 to 4 700 L/s (1 000 to 75 000 gpm). Because of structural limitations of the torque tube, the lift of the open screw pump typically is limited to 7.5 to 9 m (25 to 30 ft), but with special construction can be extended to 12 m (40 ft).

Figure 4.16 Open screw pump installation.

Standard angles of incline of an open screw pump are 30 and 38 degrees. A pump installed at 30 degrees will pump more than one installed at 38 degrees but will require more space and have less available lift.

The torque tube is supported by bearings at the bottom and the top of the pump and the drive motor and gear box are located at the top of the pump above the discharge point. The bottom bearing typically is of the grease-lubricated sleeve type and, because it is submerged, should be lubricated automatically. The top bearing is designed for both thrust and radial loads.

Open screw pumps require close clearance between the perimeter of the screw and the trough to maintain pump efficiency. When fill concrete is poured in the trough, screed boards are attached to the flight to provide the correct clearance.

Because there is clearance between the flights and the trough, the pump is self draining. This feature reduces starting loads and eliminates problems with freezing but reduces pump efficiency. Normal shaft efficiencies of open screw pumps are 75% at maximum capacity and 65% at 25% of capacity.

The open, rotating flights of the open screw pump present a safety hazard, and it is strongly recommended that a grating or cover be installed over the pump trough. In warm climates, sun screens may be required to prevent idle pumps from bowing from extreme temperatures.

Enclosed screw pumps have conically pitched helical flights welded inside an exterior steel cylinder. They are available with diameters from 0.6 to 2.75 m (24 to 110 in.) and capacities from 60 to 2 200 L/s (1 000 to 35 000 gpm). Standard angles of incline are 38 and 45 degrees. A pump installed at 38 degrees will pump more than a pump installed at 45 degrees.

The upper bearing of the pump is located above the discharge point and designed for thrust and radial loads. The lower pump cylinder is supported by steel reinforcing rings that ride on roller bearings. The lowest bearing is located above the pumped fluid with the inlet to the pump cantilevered so that the bottom bearing is not submerged.

Unlike the open screw pump, the lift of an enclosed screw pump is not restricted by the structural limitations of the torque tube. For lifts of more than 7.5 m (25 ft), additional roller bearings can be mounted on the cylinder to extend the lift.

The shaft efficiency of the enclosed screw pump is approximately 90% for flows between 50 and 100% of maximum flow and approximately 80% at 25% of maximum flow. The reason for the high efficiency is the elimination of the recycle flow of the open screw pump. The only losses in an enclosed shaft pump are bearing losses and gear box losses.

One disadvantage is that the enclosed screw impeller is not self draining, increasing the starting torque of the motor and, in cold climates, causing problems with freezing in the cylinder. To eliminate this problem, a clutch can be installed in the gear box to allow the pump, after being stopped, to go into reverse rotation to drain. The clutch limits the speed of the reverse rotation. Motion detectors are needed to prevent restart while the pump is in reverse rotation.

Another potential problem with an enclosed screw pump is the danger of air binding. If the water level at the inlet to the pump rises above the top of the exterior steel cylinder, air is trapped in the pump, severely reducing pump capacity. To prevent air binding, inlet level controls should be installed to sound an alarm and start standby pumps before the wet well level reaches the crown of the pump cylinder.

Enclosed screw construction is simplified because the trough required by the open screw pump is eliminated.

PNEUMATIC EJECTORS. Pneumatic ejectors are used where the flow of wastewater to be pumped is less than the capacity of a small centrifugal pump. The normal capacity of a pneumatic ejector is 20 L/s (320 gpm) or less.

A typical pneumatic ejector, as shown in Figure 4.17, consists of a receiver, called a pot, an air compressor, controls, and associated piping. Check valves are located on the inlet and outlet of the pot and a vent line is connected to the top of the pot. A three-way solenoid valve is installed on the vent line. In one position, the pot is vented to the atmosphere and wastewater can flow in through the inlet valve. When the pot is full, the controls reverse the solenoid valve, admitting compressed air into the pot and forcing the wastewater out through the discharge check valve. When the pot empties, the controls reverse again, venting the pot.

Because the pot is filled by gravity flow, the pot must be several feet below the inlet sewer to provide sufficient head to overcome the friction loss in the inlet pipe and check valve.

The pot typically is sized for a 1-minute cycle and the holding capacity of the pot is numerically equal to the ejector flow rate. For instance, a 6-L/s

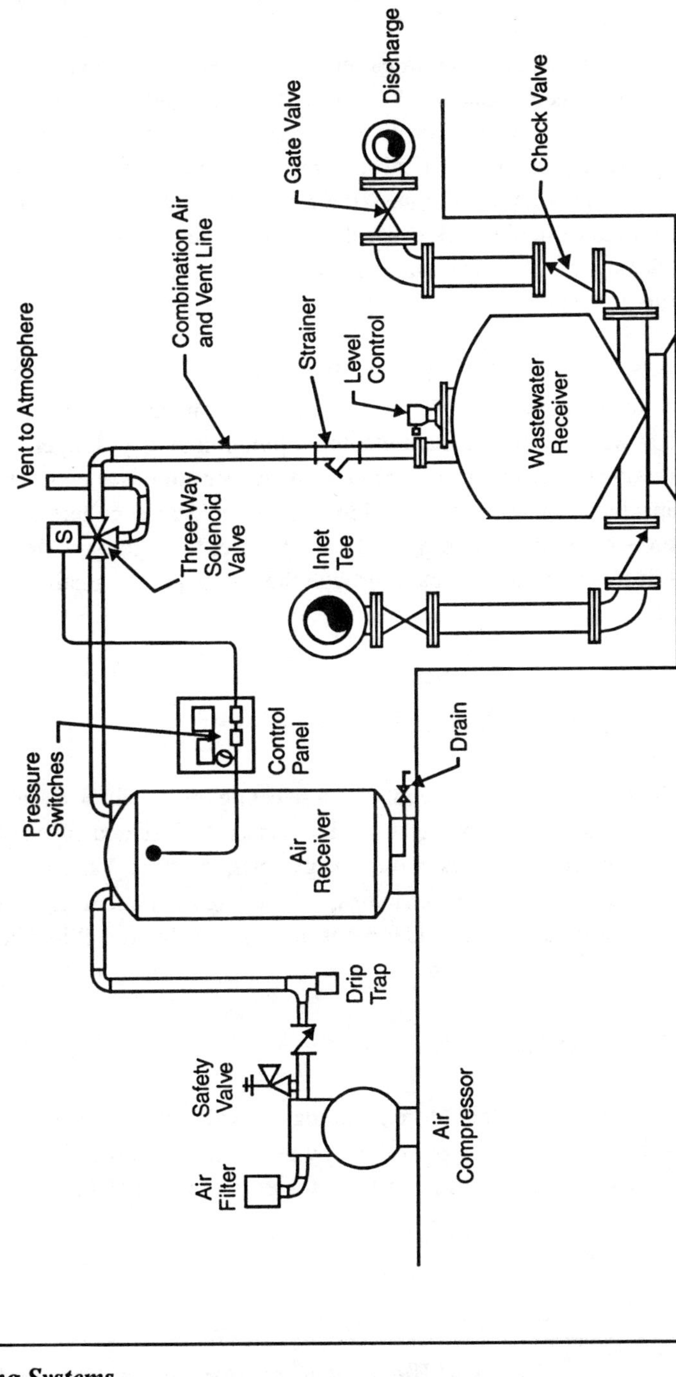

Figure 4.17 Pneumatic ejector and associated piping.

(100-gpm) ejector would have a 360-L (100-gal) capacity pot. For a 1-minute cycle, the pot fills for 30 seconds and then ejects in 30 seconds. Therefore, the inlet piping and the discharge piping and force main must be sized for 200% of the ejector rating. The instantaneous discharge rate of the 6-L/s (100-gpm) ejector would be 13 L/s (200 gpm).

Pneumatic ejectors typically are furnished in pairs so that one pot is filling while the other is ejecting. Ejectors are available with steel or cast iron pots. Cast iron reduces corrosion and is recommended for municipal service.

AIR LIFT PUMPS. Air lift pumps seldom are used in wastewater pumping stations because their total lift is 1.75 m (6 ft) or less and their efficiency is merely 20 to 30%. However, air lift pumps virtually are uncloggable.

The air lift consists of a vertical riser pipe submerged in a wet well with an air diffuser at the bottom. Compressed air is discharged through the diffuser to mix with the wastewater. Because the air and water mixture density is less than the density of the surrounding wastewater, atmospheric pressure forces the mixture to a higher head.

PROGRESSIVE CAVITY PUMPS. Progressive cavity pumps, by themselves, typically are not used for wastewater or stormwater pumping, but are common at wastewater treatment plants for pumping heavy sludges.

Progressive cavity pumps can be used in grinder pump systems. These are small-capacity pumps normally used in remote, residential collection systems. Before entering the pump, wastewater passes through a grinder to reduce solids to a size that will pass through the pump and the small-diameter force main.

FORCE MAINS

Two conflicting requirements are factors in selecting force main size. Velocity in the force main should be kept high to ensure transport of grit and other solids, yet low velocity best reduces friction loss, energy costs, and water hammer.

Requirements for sizing force mains, calculating force main friction losses, and water-hammer considerations are discussed in detail in Chapter 5.

WET WELLS

Wet wells store wastewater before pumping, provide submergence over the pump suction, and, most importantly, provide a smooth transition of flow to the pump suction. Design considerations for the wet well are discussed in Chapter 3.

PUMP DRIVES

Almost all pumps are driven by either electric motors or internal combustion engines. The most important criterion for selecting the type of drive for stormwater and wastewater pumping stations is reliability. The electric motor is more reliable and requires less maintenance than an engine drive.

Where reliable sources of power exist, electric motors are preferred. To ensure reliable power, it is good practice and often required that power be obtained from two independent power services.

Where dual power sources are not available, the best choice is direct drive of the pumps by engines or the installation of a standby engine generator. The engine generator is advantageous in that power can be furnished to any of the installed pumps and the control system and station auxiliary systems, such as lights, heating, and ventilation.

Other considerations in pump drive selection include cost of drive, operating costs, continuous versus intermittent operation, the need for variable-capacity operation, and the need for remote or automatic operation.

When evaluating the cost of the drive, expenses in addition to the first cost of the equipment must be included. These include the cost of controls, conduit, wiring, and starting equipment for motors and additional building space, extra floor strength, fuel system, jacket water cooling system, gear drives, and additional material-handling equipment for engines.

Operating costs include energy or fuel maintenance material and personnel expenses and the cost of operating auxiliary systems.

The duty cycle of the station can affect the energy costs of a pumping station. Electrical power charges usually have two components, the energy charge and the demand charge. The energy charge is based on total energy used, and the demand charge is based on highest use of power over a given period of time. The cost of fuel used in engines does not have a demand charge, favoring their use at stormwater pumping stations, which operate intermittently so use little energy. When operating, however, their energy demands are high.

An engine does not have restrictions on time between starts, but the frequency of starting a motor is restricted because of the heat generated by the inrush current when starting. The time required between starts is discussed in Chapter 3.

Engines are designed for variable-speed operation. Electric motors are constant-speed drives and require additional controls or equipment to adapt them for variable-speed operation.

Electric motors can easily be adapted to automatic control and remote operation. The automatic control of an engine is more complicated, as is monitoring from a remote location because of engine auxiliary systems.

ELECTRIC MOTORS. The following is an overview of motor drive requirements when applied to pumping systems. Electric motors are designed for either alternating current or direct current. Direct current motors seldom are used in municipal systems, where they would only be used in variable-speed systems.

Alternating-current motors can be either single phase or three phase. The motor rating of single-phase motors is limited to approximately 7.5 kW (10 hp), so almost all motors used in pumping stations are three-phase, alternating-current motors.

The three types of alternating-current motors are the squirrel cage induction motor, the synchronous motor, and the wound-rotor induction motor. The speeds of the synchronous motor and the squirrel cage induction motor are constant. Wound-rotor induction motors also operate at a constant speed but can be made to operate at different speeds by varying the resistance in the motor secondary circuit.

The synchronous speed of a motor is calculated by the following equation:

$$\text{r/min (rpm)} = \frac{120 \times \text{frequency}}{\text{Number of motor poles}} \qquad (16)$$

Synchronous motors operate at this speed but squirrel cage and wound-rotor induction motors at full speed have a 2 to 3% slip loss and, therefore, operate at 97 to 98% of the synchronous speed.

The speed of stormwater and wastewater pumps should not exceed 1 800 r/min (rpm), and 1 800-r/min pumps should be used only when the system head requires a pump of this speed. Most stormwater and wastewater pumps operate in the range of 1 200 to 600 r/min to reduce wear on the pump.

Higher speed pumps are less expensive, but lower speed pumps are larger, more durable, and less susceptible to wear and vibration. As pump size increases, pump speed decreases. Pump head is proportional to the square of the tip speed of the impeller. A larger diameter impeller will rotate at a lower speed to develop the same tip speed of a smaller impeller operating at a higher speed.

Squirrel cage induction motors are used most often because they are low in initial cost, require little maintenance, and have a proven record of reliability. However, at speeds below 600 r/min, squirrel cage induction motors have low power factors, which may increase power costs.

Synchronous motors are more expensive than squirrel cage induction motors and typically are used only in the larger sizes (400 kW, or 530 hp) and at lower speeds. However, because the synchronous motor can operate at unity or a leading power factor, it can be used in systems where power factor correction is required or it is beneficial to reduce power costs.

Wound-rotor motors are more expensive than squirrel cage induction motors and require more maintenance because of the brushes on the motor slip rings but can be made to operate at speeds below full speed.

Pump-system operation is often improved and energy costs are reduced if two-speed motors or step-speed control of wound-rotor motors is used.

The wound-rotor motor operates at peak efficiency only when operating at its maximum speed. At reduced speeds, the motor is subject to slip losses generated by the resistance in the wound-rotor motor secondary circuit. Slip losses can be calculated by the following formula:

$$\text{Slip loss} = \text{Pump power} \times \frac{(\text{Full speed} - \text{Reduced speed})}{\text{Reduced speed}} \qquad (17)$$

Where

Slip loss	=	power lost, kW (hp), and
Pump power	=	pump shaft power at the reduced speed, kW (hp).

The total power supplied by the motor is the pump shaft input power plus the slip loss. The energy in the slip loss normally is released in the form of heat in a resistance bank, which in cold climates can be used to heat the building. In warmer climates, or during warm periods, this heat must be wasted.

An additional advantage of the step-speed control of the wound-rotor motor is its ability to use the steps to bring a motor up to full speed slowly. This "soft start" eliminates the high inrush current of a squirrel cage induction motor and reduces the required time between starts.

Squirrel cage induction motors are available for operation at two different speeds. However, unlike the wound-rotor motor that can be adjusted to operate at any speed, the lower speed of the two-speed induction motor must be a synchronous speed less slip loss. Two-speed squirrel cage induction motors can be two-speed, two-winding motors or two-speed, one-winding motors.

The ratio between speeds of a two-speed, two-winding motor can be the ratio between any two synchronous speeds. The reduced speed of a two-speed, one-winding motor must be 50% of the higher speed. Unlike the wound-rotor motor that loses efficiency at reduced speed, the two-speed squirrel cage induction motor maintains high efficiency at both speeds because the two-speed motor is, in fact, two separate motors.

VARIABLE-SPEED MOTOR OPERATION. Pump systems that must match inflow to the station normally are furnished with variable-speed pumps. Electric motors basically are constant-speed drives. There are, however, ways to provide stepless, variable-speed control with electric motor drives.

Variable-frequency Drive. The variable-frequency drive is the most common method of variable-speed drive, generally replacing the older forms of variable-speed drives. This drive has been in common use since the early 1980s and is an efficient, reliable method for variable-speed operation of squirrel cage induction motors, and, in larger sizes, synchronous motors.

The variable-frequency drive takes alternating current, converts the power to direct current, and then reconverts the direct current to alternating current at a different frequency. Because the synchronous speed of a motor is proportional to frequency, the speed of the motor depends on the output frequency of the variable-frequency drive controller. Because the motor is operating at its full synchronous speed, there are no slip losses in the control.

There are, however, losses in the controller for converting to direct current and reconverting to alternating current. These losses are fairly constant over the full speed range of the controller. Therefore, the variable-frequency drive has poorer efficiency between 95 and 100% of full speed than other variable-speed drives that have slip losses, but the efficiency at less than 95% speed is much better than other drives.

There are several types of variable-frequency drives, with the pulse width modulation and current source converters being the most prevalent. The output power produced by the variable-frequency drive is not a smooth sinusoidal wave and induces harmonic disturbances and a low-power factor on the power supply source. Adding reactance to the circuit can help but does not eliminate the problem. For this reason, motors often are sized to operate at 80 to 90% of their rated power and are furnished with a 1.15 service factor.

Care must be taken in the selection of engine–generator sets for use with variable-frequency drives. In addition to the 5% power loss in the converter, some oversizing is required to compensate for the harmonics in the power supply, especially if two or more variable-frequency drives are powered by the generator.

Variable-frequency drives can be furnished to operate several motors in parallel. This reduces the initial cost of the variable-speed drive but is not a recommended practice because of the need for reliability. If the variable-frequency controller malfunctions, it would shut down too large a portion of the pumping station. A separate variable-speed controller is recommended for each variable-speed pump.

Other methods of producing stepless, variable-speed electric drives are available in two basic forms: (1) where the output speed of the motor is variable and (2) where the output speed of the motor is constant but operates through a clutch that produces a variable-output speed.

Wound-rotor Motor Variable Speed. The output speed of wound-rotor motors can be made stepless if the resistance is infinitely variable.

A solid-state controller is available that converts the energy in the slip losses of the secondary circuit to direct current, reconverts the direct current to alternating current, and feeds this power back into the power supply to the drive. This control is known as wound-rotor regenerative secondary control.

The regenerative secondary controller efficiency approaches that of variable-frequency drives in that the energy in the slip losses is converted to useful work. However, the regenerative secondary controller has not been popular because of the additional cost and maintenance requirements of wound-rotor motors. They do, however, have energy-saving potential on existing systems with wound-rotor motors.

Liquid rheostats have been used for many years with wound-rotor motors to provide stepless-speed control. In these controls, the resistance in the secondary is varied by changing the area of electrodes covered by an electrolyte.

It is important that electrolyte strengths be maintained equally in rheostats that control pumps that operate in parallel. If electrolyte strengths differ, pump speeds will be different because, even though the submerged area is equal, the secondary resistance will be different.

The energy of the slip losses heats the electrolyte, which then must be cooled. This heat can be used to heat the building by passing the electrolyte through a heating coil, or the heat can be transmitted to the fluid being pumped to liquid heat exchangers on the pump suction or discharge piping or can be exhausted to the outdoors.

Variable-voltage Motors. Variable-voltage motors are specially designed high-slip squirrel cage induction motors. The primary voltage to these motors is varied by saturable reactors or silicon-controlled rectifiers to vary the output speed of the motor.

This type of control is not recommended for municipal service because it is inefficient and slip losses generate heat in the motor, causing it to run hot.

Variable-speed Clutches. Variable-speed clutches have been available for years but are less common today because of the development of the variable-frequency drive. Clutches are available in magnetic drive and fluid drive. A squirrel cage induction motor most often is used to drive both clutches.

The magnetic drive, or eddy-current coupling, consists of a steel ring member, or drum, driven at constant speed by the output shaft of the motor and magnetic coils mounted on the output shaft of the drive. The strength of the magnetic field of the coils is varied to adjust the slip speed between the input and output shafts. These drives generate slip losses, with the lost energy calculated by equation 17. In addition to the slip losses, bearing and windage loss must be included in calculating total energy loss.

Most magnetic drives are air cooled with heat generated by the losses exhausted to the space around the drive. This heat can be used for space heating, but during warm weather must be exhausted. Large units can be water cooled.

Hydraulic drives are similar to magnetic drives with the slip between input and output shafts controlled by the viscous shear of oil that separates the drive and the driven clutch plates.

Other types of variable-speed drives, such as adjustable belt pulleys and shift gears, are not used in wastewater and stormwater pumping stations. Figure 4.18 shows the relative efficiencies of the various types of variable-speed drives.

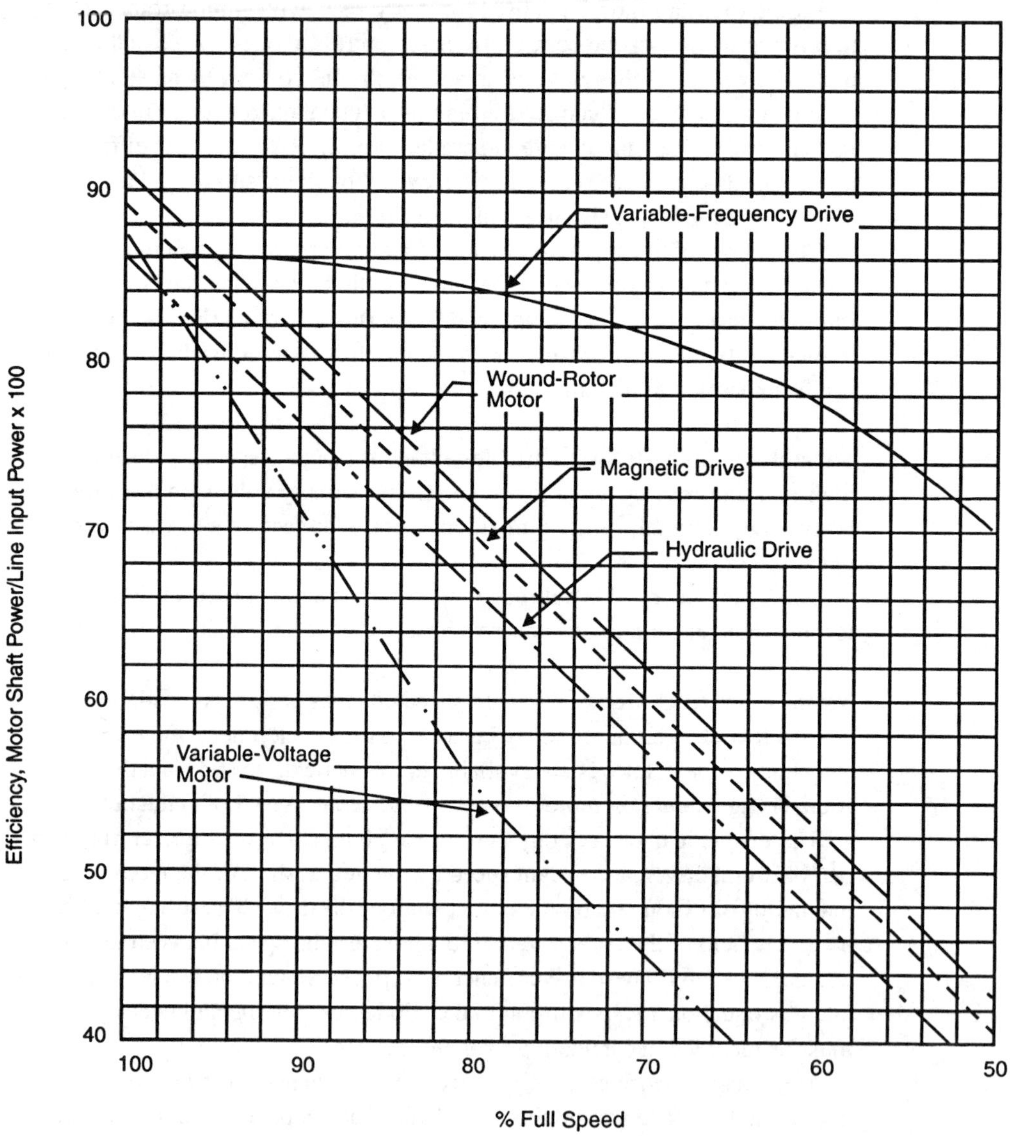

Figure 4.18 Variable-speed drive efficiencies (approximate efficiencies; actual efficiency will vary with size and speed of motor and system curve characteristics).

INTERNAL COMBUSTION ENGINES. Internal combustion engines can be fueled by gasoline, diesel oil, natural gas, liquid petroleum gas, or, if available, sludge gas. Engines can be spark ignited, with the fuel ignited by a spark plug, or diesel engine, with the fuel ignited by the heat of compression inside the cylinder. Dual fuel engines are available with gas and diesel oil that are ignited by the heat of compression of the diesel fuel.

Most engine drives include a gear box between the engine and pump to allow both pieces of equipment to operate at their most efficient speed. Right-angle gears are used where the engine is horizontal and the pump is vertical.

Low-speed, spark-ignited gas or diesel engines are most often used and, with proper maintenance, will operate for years. High-speed gasoline engines are the least expensive but have higher maintenance costs. Also, because gasoline is more volatile than other fuels, safety is a problem and difficulties can arise with the fuel if it is stored for an appreciable time.

Engines can be naturally aspirated or pressure charged. The volume of the air–fuel mixture that can be introduced to the cylinders in a naturally aspirated engine is limited, so the power generated also is limited. In a pressure-charged engine, the combustion air is pressurized by an exhaust-driven or power-train-driven blower to compress the combustion air. This forces a larger volume of air–fuel mixture into the cylinders and increases the power and efficiency of the engine. Also, the exhaust gas from a pressure-charged engine tends to be cleaner than the exhaust gas from a naturally aspirated engine.

One advantage of engine drive is that the engine does have the capability of variable-speed drive. Occasionally, an engine can be used to provide standby drive for a pump by mounting the motor on the top of a dual right-angle gear drive with the motor shaft connected to the gear output shaft and the engine connected to the horizontal gear input shaft through a clutch that disconnects the engine when the motor is used. When the engine drives, the motor rotor windmills.

Auxiliary Systems. Auxiliary systems required for internal combustion engines include the fuel system, lubrication system, jacket water cooling system, and exhaust system.

Stations designed for continuous operation that use gasoline or diesel fuel must have sufficient storage space to receive the full capacity of delivery trucks plus several days' reserve in the event delivery is delayed.

Liquid petroleum gas (LPG) occasionally is used in small systems or where an engine is used for standby power. Though stored under pressure as a liquid, LPG is used in the engine as a gas. It can be withdrawn from storage as a liquid or a gas, but if large quantities are withdrawn as a gas, the heat of vaporization cools the liquid to prevent further gas withdrawal. Therefore, LPG typically is withdrawn as a liquid and vaporized to a gas in a vaporizer, using heat available in the jacket water.

Natural gas typically is piped to the station from a gas distribution system, which should have a reliable gas supply and adequate pressure. Turbo-charged and intercooled engines require gas at high pressures [170 to 210 kPa (25 to 30 psi)] but are more efficient than naturally aspirated engines.

Only 25 to 35% of the energy supplied to an engine is converted to shaft output power. The balance of the heat is transmitted to the jacket water and the exhaust gas with a small amount (6 to 10%) radiated to the surrounding space.

Twenty to thirty percent of the engine input energy is transmitted by the engine to the jacket water. The jacket water leaves the engine at 80 to 90°C (180 to 200°F) and must be returned to the engine 8 to 10°C (15 to 20°F) cooler to cool the engine. If there is a need for waste heat, such as building heating, this is an ideal source of heat. If the heat is not required, it must be dissipated. Air-cooled radiators most often are used in smaller systems. Larger systems can use remote, mounted fluid coolers or cooling towers with jacket water–heat exchangers. Water from a cooling tower should not be used directly in the cooling system or solids in the cooling tower water might plug small openings in the engine.

Additional waste heat can be recovered from the exhaust gas by the use of heat recovery silencers. Exhaust silencers are required on all engines to prevent noise pollution.

PUMP SELECTION

Before selecting pumps, the design engineer must evaluate the system characteristics and determine the requirements for the pumping systems, which will vary with each system.

SYSTEM DESIGN CONSIDERATIONS. Pumps installed in a station must be capable of discharging the peak design flow at the system head required with aged pipe. It is normal practice, and typically required by code, to provide this pumping capacity with the largest pump out of service. In large stations, especially those with internal combustion engine drives, it may be desirable to discharge the design peak flow with one standby pump and one out of service for maintenance.

Pump selection evaluation should consider flow rates, system head-capacity curves, station location and area served, and force main velocities.

Flow Rates. The system must be designed to discharge the design peak flow, which is the maximum flow the station will be required to discharge during the design life of the station. This flow may not develop until the end of the design life of the station, and the pumps will operate at this flow for only short periods of time.

Because the design flow rate is only occasionally achieved, other flow rates are more important to the design engineer. The station should be designed to operate efficiently during initial, interim, and design year average flows.

Interim peak flow rates are the peak flows that can develop before the design life of the station is reached. Because the design life of the station can be 30 years or more, it is reasonable to install pumping equipment sized to pump the peak flow expected in the first 10 to 15 years of the station.

When the interim peak flow is reached, the station must be expanded to discharge the design peak flow. The system capacity can be increased by installing additional pumps or increasing the capacity of existing pumps.

Pump capacity can be increased by changing pump speed or increasing the diameter of the impellers installed in the existing pumps. Changing pump speed probably will require new drives. Installing larger impellers will increase the required input power to the pump, which may require new drives, or the original drives may be oversized to allow for future power requirements.

The initial minimum flow may determine the pump capacity if variable-speed pumps are installed. The lead pump must be variable speed and operate within the pump operating envelope (see Figure 4.11). At minimum flow, the pump will operate at minimum speed and flow in the operating envelope. At full speed, the pump must also operate within the pump operating range. This may limit the maximum capacity of the pump, especially if the system has a flat system head curve.

System Head-capacity Curve. The system head-capacity curve presents a graph of the system into which the pumps will operate. Details for the preparation of the system curve are given in Chapter 5.

When the system head-capacity curve has been developed, the initial, interim, and design year minimum, average, and peak flow rates should be added to help visualize the system requirements. Also, the force main velocities should be included on the curve.

Figure 4.19 is typical of a system head-capacity curve and shows a system curve for a collection system with a design population of 41 000 people and a 4 200-m long, 0.6-m in diameter (13 800 ft long, 24 in. in diameter) force main. The interim design population is 32 000 people, and the system static head is 9.1 m (30 ft). The curve includes the friction loss for new pipe and the design friction loss for old pipe and shows the flow rates and force main velocities. The velocity in the 0.6-m (24-in.) force main varies from 0.17 m/s (0.6 ft/sec) at minimum flow to 1.7 m/s (5.5 ft/sec) at peak flow. If a 0.75-m (30-in.) force main were used to reduce friction loss, the velocities would be reduced to 0.1 m/s (0.33 ft/sec) at minimum flow and 1 m/s (3.3 ft/sec) at peak flow. These low velocities could cause problems with settlement of solids.

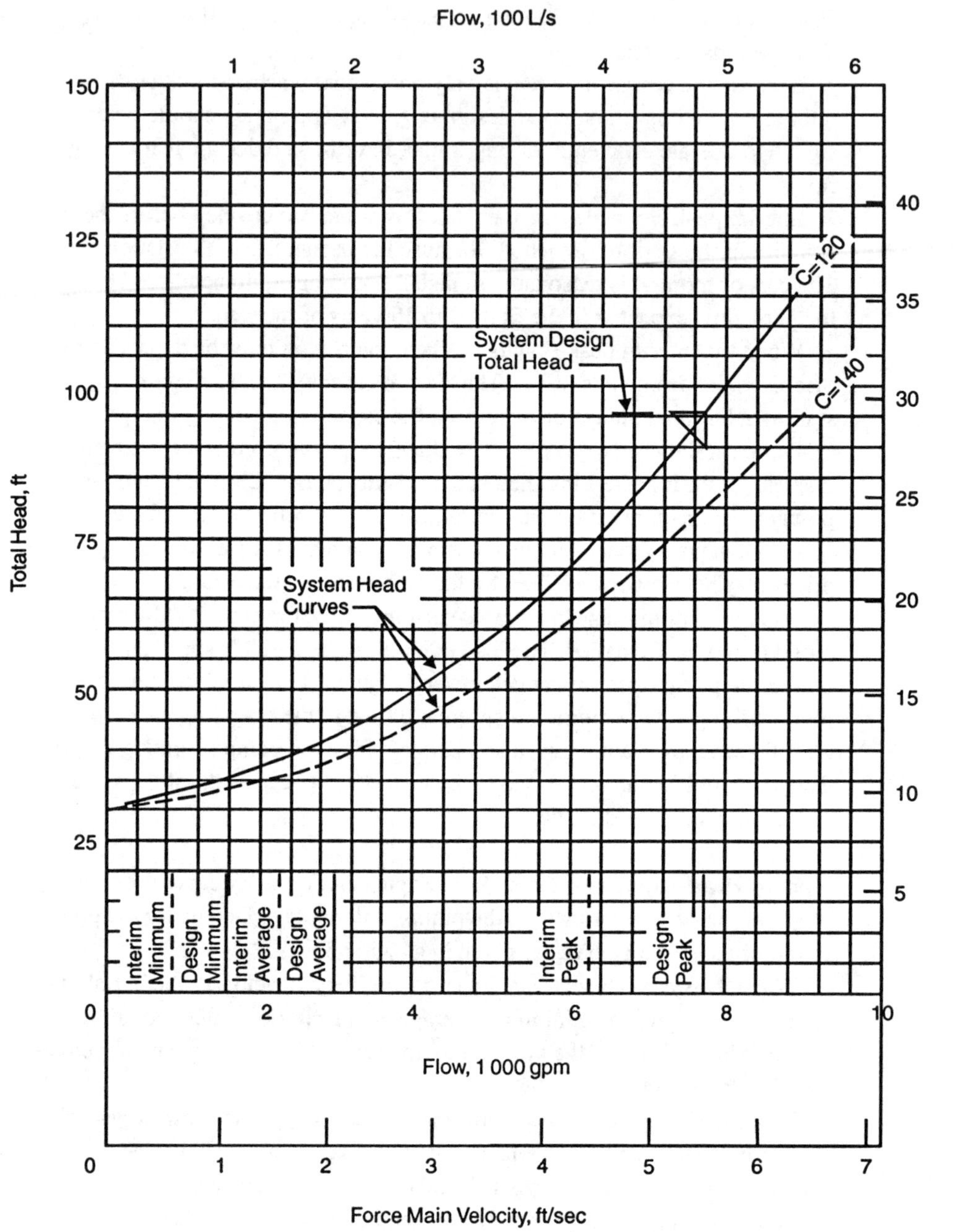

Figure 4.19 Typical system head-capacity curve (ft/sec × 0.304 8 = m/s).

Station Location and Area Served. The station location and the area served may determine the type of pump control (on–off or variable capacity), the need for future growth, and the type of pump drive.

Stations that serve only a small portion of a collection system typically will be furnished with on–off pump control or two-speed pumps. On–off pump controls are the simplest types, but two-speed pumps may discharge the flow with lower power costs.

If the station is located in an area that has potential for substantial future growth, the design peak flow and initial peak flow must be studied with care to ensure efficient operation. An oversized pumping system will use more energy than one designed for realistic flows. Consideration should be given to designing the station for interim peak flows, with the potential to increase station capacity in the future. However, this may affect force main sizing.

Pumping stations that discharge to a treatment plant or large stations that discharge a large portion of the total collection system flow should be furnished with variable-speed pumps or step-speed control with small step increments. This will prevent sudden changes in plant flow that could upset the treatment process.

Force Main. Pump rates in the system must provide force main velocities that will both reduce settlement of solids in the force main and resuspend settled solids.

SELECTION PROCEDURE. The recommended steps before selecting pumps for use in a pumping system include preparation and analysis of the system head-capacity curve, selection of the number of pumps, and preparation and analysis of system curves.

System Head-capacity Curve Characteristics. The characteristics of the system head-capacity curve play an important part in the selection of the pumping equipment. The system curve is required to determine the rating of the pumps and is needed to determine the number of pumps in a system.

A system head-capacity curve with a high friction loss component and a small static head component favors the use of a single pump with two speeds or variable-speed drive. By the affinity laws, the pump head decreases in proportion to the square of the speed ratio and the flow in proportion to the speed ratio. The system friction loss, however, is proportional to the flow reduction ratio squared. Therefore, if a single pump is selected to discharge the design flow at its best efficiency point (BEP), a pump operating at reduced speed also will operate near its BEP.

A system with a large static component and a small friction component favors multiple pumps because if a pump is selected that, at full speed, operates near run-out on its operating range, a small reduction in speed would operate the pump at the minimum flow in the operating range envelope.

This is illustrated in Figure 4.20, which shows a high friction component system curve and a high static component curve superimposed on the variable-speed pump curve shown in Figure 4.11. The speed range and the flow range for the high friction component is larger than the system with a high static component and the efficiency remains fairly constant. The efficiency of the system with the high static component goes from one extreme to the other of the operating envelope.

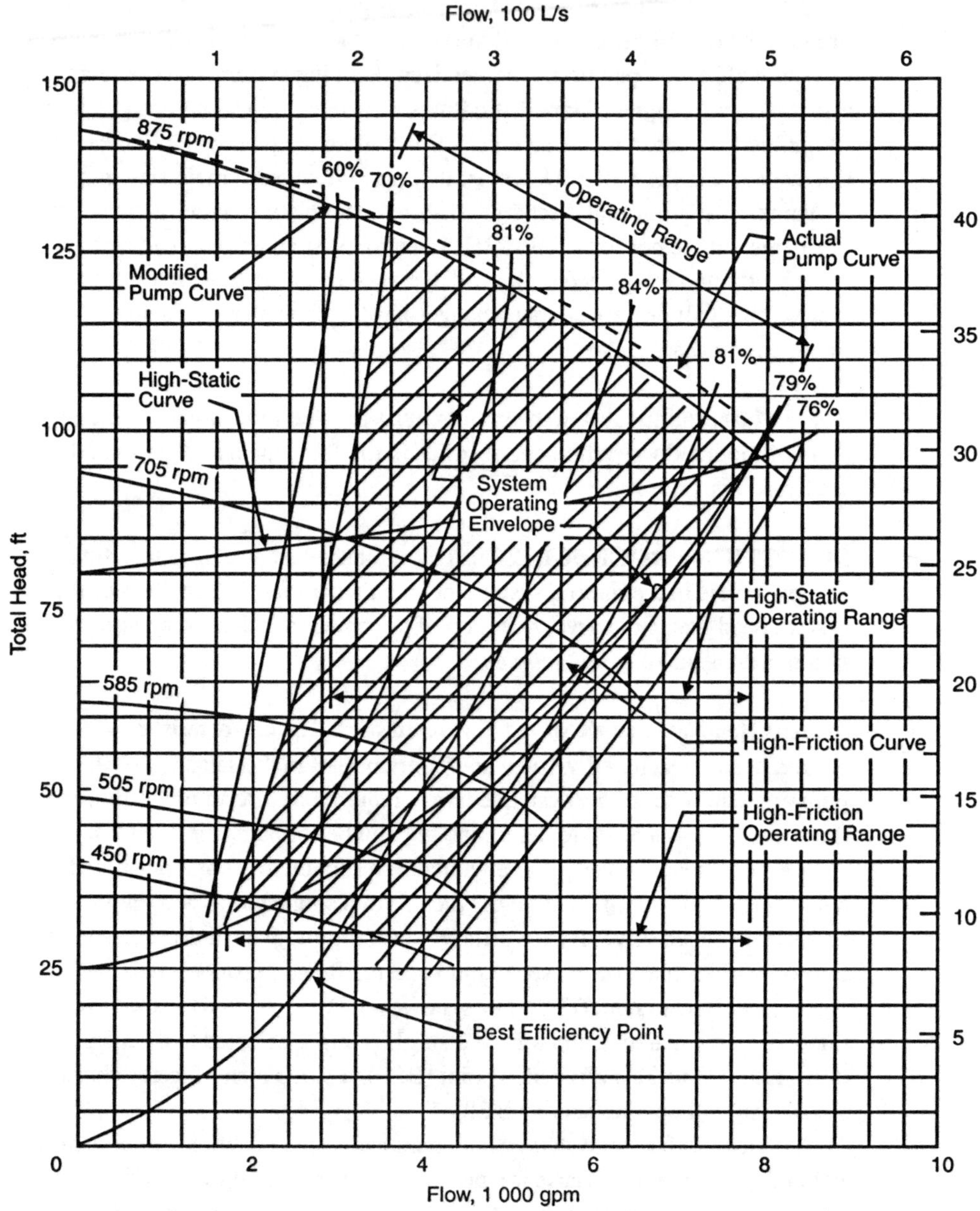

Figure 4.20 Effect of system head-capacity curve on pump operation.

Preliminary Selection of Pumps. Preliminary pump selection can be made after the initial screening of station requirements. Because most municipal pumping stations require competitive bidding, pumps from several manufacturers should be included in the preliminary selection.

In stations with multiple pumps, the pumps can have equal or different capacities. System operation with different-capacity pumps is similar to multi-speed operation of a pump system. When the inflow exceeds the capacity of the small pump, the larger pump would start and the smaller pump would shut off. As flow increases, the smaller pump restarts. This operation, however, requires frequent starting and stopping of pumps, and the wet well must be sized for these capacity steps.

In stations requiring stepless flow control, the variable capacity of centrifugal pumps can be provided by variable-speed operation of the pump or by throttling the discharge of a constant-speed pump.

Variable-speed operation is the preferred method of capacity control and will provide a larger turndown in pumped capacity than pump throttling, which only reduces pump flow rate and is limited by the allowable operating range. Capacity control by speed reduction reduces the pump flow and the discharge head as speed is reduced.

When a pump capacity is varied by throttling, the pump operates at higher heads than necessary and, at reduced flows, with poor efficiencies. Also, a valve used for throttling operates with a large pressure differential and is subject to clogging and wear. Because of the wear, the throttling valve should not be used for isolation or check valve service.

Caution is required for pump systems with variable-capacity pumps to ensure pump operation remains within the operating range envelope.

In stations designed for variable-capacity operation with equal-capacity pumps, the lead and first lag pumps should have variable capacity. Additional pumps, if required, can be constant speed. If the first lag pump were constant speed, the lead pump would have to operate outside the pump operating envelope for flows slightly greater than the capacity of the first lag pump.

If the lead and first lag pumps have different capacities, the larger pump should be in the lead position and have variable capacity. When the smaller, constant-capacity pump starts in the first lag position, the lead pump can operate at reduced capacity but still within the pump operating range.

In stations with variable-speed pumps, the standby pump should also be a variable-speed pump. The lead pump and first lag pump will do most of the work, with the second lag pump periodically operating. Therefore, the standby pump should be variable speed to provide reliable operation and share the station work load. In large stations requiring three or more operating pumps, variable-speed capability should be considered for most, or all, pumps.

Preparation of System Curves. Following the preliminary selection of pumps, prepare a system curve for each of the preliminary pump systems.

A system curve is the combination of the system head-capacity curve and the pump curve. This system curve presents a picture of how the system will operate. A pump will discharge the flow where the pump curve intersects the system head-capacity curve. The system curve should be drawn only for the piping common to all pumps, so station losses should not be included in the system head-capacity curve.

Station losses are the friction losses in the suction and discharge piping of an individual pump and are deducted from the pump head-capacity curve to develop a "modified pump curve" for the individual pump. This is the curve the system curve will "see" the pump producing. The station losses must be added to the system head to determine the pump total head. Figure 4.21 illustrates a modified pump curve in which the station losses have been deducted from the actual pump head-capacity curve.

The reason for using the modified pump curve is to simplify the analysis. In a two-pump system, each pump discharges 50% of the total flow. If station losses were included in the system curve without recognizing that each pump handles only 50% of the flow, station losses would be calculated at four times the actual loss.

Modified curves can be used in developing the pump operating envelope. Once the full-speed modified curve has been developed, it can be used with the affinity laws to predict the pump curve at reduced speeds because the pump speed at reduced flow is proportional to the speed ratio and the head proportional to the square of the speed ratio.

Correspondingly, station losses at reduced speed and, therefore, flow, are proportional to the square of the flow ratio and the square of the speed ratio.

Analyzing System Curves. Candidate system curves must be analyzed to allow selection of the optimum system. The following requirements should be considered:

- System $NPSH_A$ is greater than pump $NPSH_R$ under all operating conditions;
- Pump operation is within the pump operating envelope;
- Pump operation is efficient at interim and design average flows;
- Energy consumption is based on expected diurnal flow patterns;
- Wet well volume is adequate to prevent short cycling of constant-speed pumps and smooth operation of variable-speed pumps; and
- Force main velocities are within acceptable limits.

EXAMPLES OF PUMP SELECTION. The following section presents different types of system curves with pumps selected for operation with these curves, illustrates some of the correct applications of pumps to a system curve, and points out several potential misapplications.

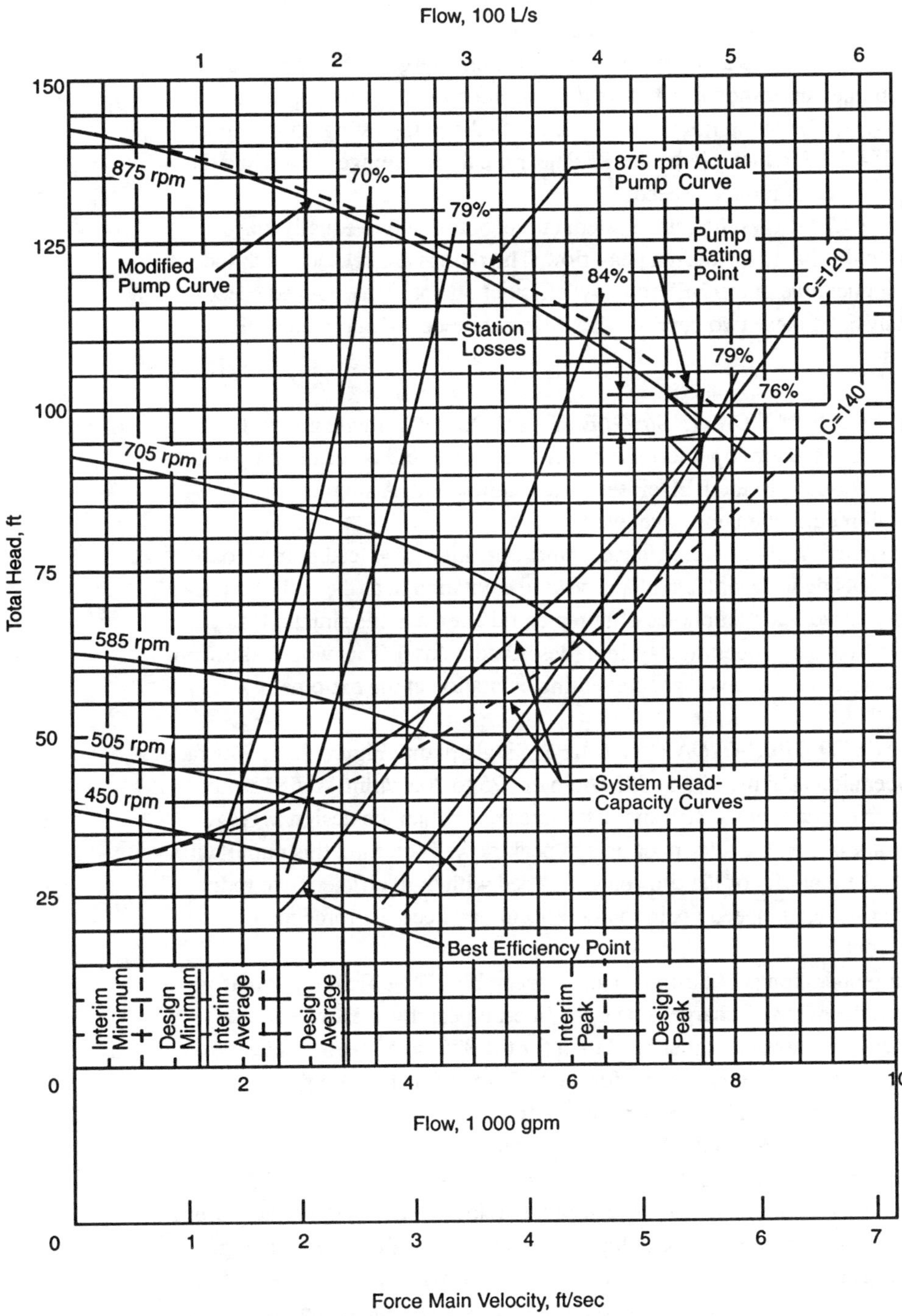

Figure 4.21 High friction head system curve, one pump (ft/sec × 0.304 8 = m/s).

High Friction Head System. The system head-capacity curve shown in Figure 4.19 is typical of a system with a small static head component and a large friction head component. The peak flow can be pumped by either one large pump or two smaller pumps operating in parallel.

Figure 4.21 shows a system curve with a single large pump and is based on the pump curve shown in Figure 4.8 and the operating envelope shown in Figure 4.11. The pump has been selected with the pump rating point slightly beyond its BEP, which places the BEP near the daily average flow when the pump operates at reduced speed.

Figure 4.22 shows the same system curve, but with two equally sized pumps to discharge the design peak flow. The pumps are selected with the rating point near the pump BEP to allow efficient operation at reduced speeds. An analysis of these two curves, with different modes of pump control, follows.

ON–OFF OPERATION OF SINGLE-SPEED PUMPS. Single pump, Figure 4.21. Satisfactory operation. A better selection could be made if a pump could be found with the BEP closer to the rating point. A single pump, however, will require the largest wet well.

Two pumps, Figure 4.22. Unacceptable operation. The lead pump would operate outside its design envelope, risking cavitation near the end of its head-capacity curve. A better choice of pump could alleviate this situation, as would a two-speed operation system (described below). This would allow operation closer to the envelope, preventing cavitation at the extreme.

TWO-SPEED OPERATION OF PUMPS. Single pump, Figure 4.21. Satisfactory operation at either 875 to 585 r/min or 875 to 705 r/min. At 585 r/min, the pump operates at its BEP and discharges more than the design average flow but less than the interim or design peak flow. Therefore, this pump will operate the majority of the time on low speed with high efficiency to reduce power costs. High-speed operation will occur and keep the force main flushed out.

Pump operation at 705 r/min also is acceptable. The efficiency is reasonable and the pump discharge is near the interim peak flow. However, the pump will operate at a higher head than for the 875- to 585-r/min selection, and the total power consumption for the 875- to 705-r/min pump will be greater than for the 875- to 585-r/min pump.

Two pumps, Figure 4.22. Satisfactory operation for 1 175 to 875 r/min, but the controls must be set to eliminate operation of one pump at high speed. The control conditions would be lead pump at low speed, lead and lag pumps at low speed, and both pumps at high speed. The step of one pump at high speed plus one pump at low speed should be used with caution because the pump operating at low speed would be functioning outside its operating envelope, near shutoff head.

Pump speed of 1 175 to 705 r/min should not be used because the change in flow between two pumps at low speed and two pumps at high speed is too great.

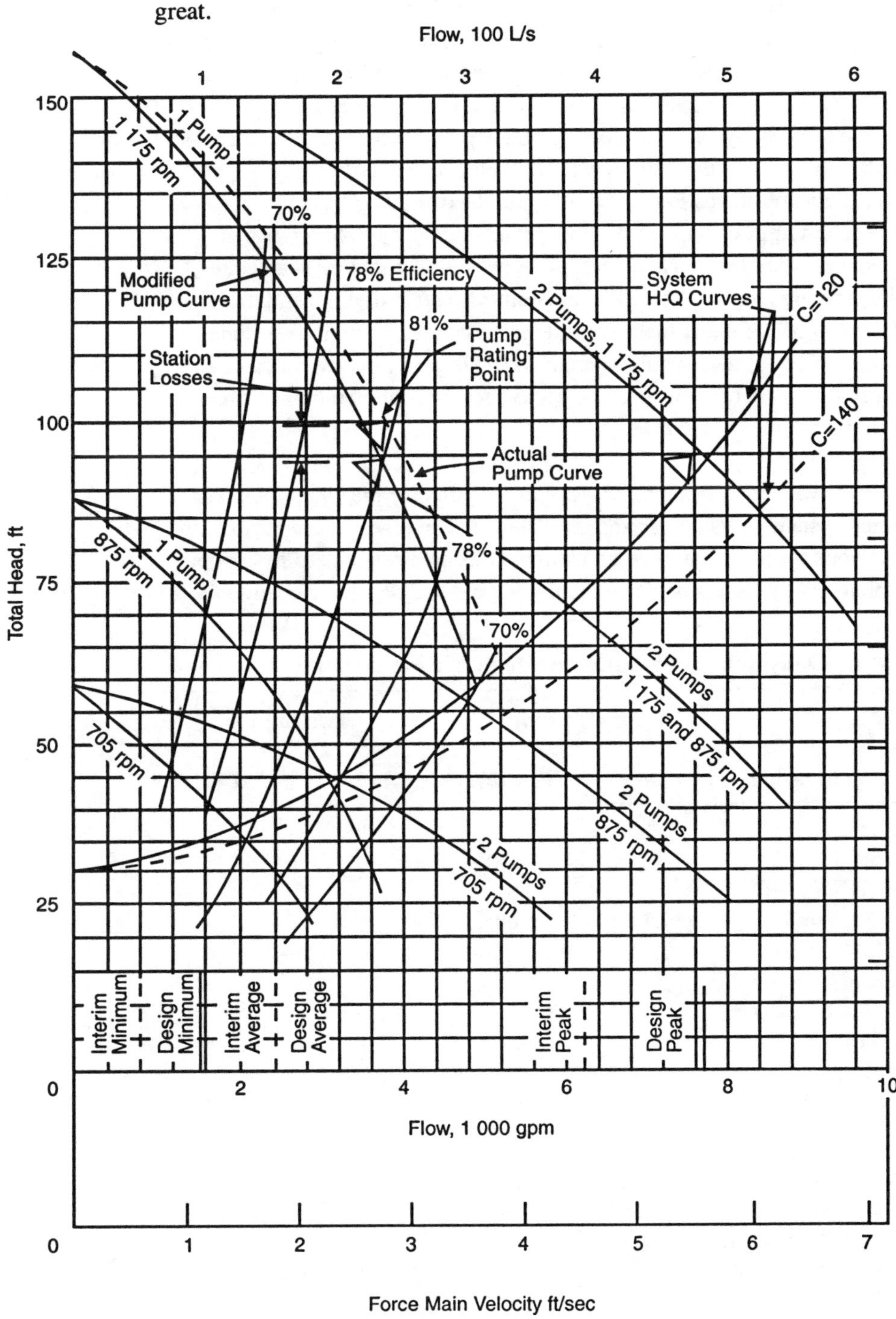

Figure 4.22 High friction head system curve, two pumps (ft/sec × 0.304 8 = m/s).

VARIABLE-CAPACITY OPERATION OF PUMPS. Variable capacity by throttling is not acceptable for either the one- or two-pump system. The pumps can only be throttled to 225 L/s (3 600 gpm) for the single pump and 150 L/s (2 400 gpm) for the two-pump system without straying from their operating envelopes. These flows are at or greater than the system average flow. Also, throttling would require more energy than variable-speed pumps.

Variable capacity by variable-speed operation provides a wider flow variation than by throttling. The minimum flow by variable speed is 110 L/s (1 750 gpm) for the single pump and 65 L/s (1 000 gpm) for the two-pump system.

The single-pump system would operate outside the operating range at minimum flows but this may be acceptable because the system head is low, reducing bearing loads and potential erosion from recirculation, and the pump would operate under these conditions for only short periods of time.

The two-pump system would operate adequately, but to prevent cavitation in the lead pump the controls must be set to start the second pump before the lead pump reaches full speed.

POWER CONSUMPTION. The power consumption for the single-pump and two-pump systems are summarized in Table 4.4. The time required for pumping at various speeds is based on an assumed diurnal flow pattern for a population of 36 000, which is between the initial year and design year flows.

Table 4.4 Daily power consumption—high friction system.[a]

	Single pump		Two pumps	
System operation	**h/d**	**kWh/d[b]**	**h/d**	**kWh/d**
Single speed				
(on–off)	7.8	1 550[c]	[d]	
Two speed				
(705/875 r/min)	9.5/0.6	1 085	—	
(585/875 r/min)	10.4/2.1	950	—	
Two speed				
(875/2 at 875/2 at 1 175 r/min)	—	—	11.1/2.6/ 1.4	900
(705/2 at 705/2 at 1 175 r/min)	—	—	12.3/3.3/ 3.0	950
Variable speed	24	1 040	24	920

[a] Power consumption based on an assumed diurnal flow variation for a system with a population of 36 000.
[b] kWh/d × 41.67 = W.
[c] Power consumption for a different single-speed pump operating at 85% efficiency at the rating point is 1 440 kWh/d.
[d] On–off operation for two pumps at full speed is not recommended because lead pump operates beyond the pump operating envelope.

The maximum power draw of 65 000 W (1 550 kWh/d) is required by a single, constant-speed pump because all flow must be pumped against the maximum system head.

Considerable power can be saved if two-speed pumps are used because the majority of flow is pumped at lower heads. The single pump operating at 585 to 875 r/min draws only 40 000 W (950 kWh/d), and the two-pump, two-speed motor operating at 875 to 1 175 r/min draws only 37 500 W (900 kWh/d). However, the two-pump system would require a larger building than the single pump.

More power is required for variable-speed operation than for the most efficient two-speed operation because of losses in the variable-speed drive and constant operation of the pumps.

High Static Head System. Systems with high static head components and small friction head components are shown in Figures 4.23, 4.24, 4.25, and 4.26. These curves illustrate the difference between pump selection for a high friction system and a high static head system. The flow rates for these curves are the same as those used for the high friction curve systems in Figures 4.21 and 4.22 and illustrate the requirements for one-, two-, and three-pump systems.

The pump in Figure 4.23 for single-pump operation is the same one used for single-pump operation as a high friction component system. This is a good selection for two-speed operation; but, for single-speed operation, a slightly larger pump is available with its best efficiency at the rating point. This larger pump cannot be used with two-speed operation because it would operate close to shutoff at low speed.

The pumps used in the two-pump system shown in Figure 4.24 are the same ones used in Figure 4.22 for the high friction head system. The BEP is at the rating point, and the pump operates efficiently when only one pump is in use.

Different pumps should be selected for a three-pump system; one for on–off operation and a different selection for a system with variable-speed operation.

For on–off operation of a three-pump system, pumps should have a BEP beyond the rating point and, if possible, near the lead-pump operation point. Because the lead pump discharges the design average flow, the system is highly efficient. When two pumps are required, the system efficiency would still be good. The least efficient system is with three pumps operating to discharge the design peak flow, but this occurs only for short periods of time. Figure 4.25 illustrates three pumps selected for on–off operation.

The pumps selected for variable-speed operation in a three-pump system should have their best efficiency near the pump rating point and, at full speed, the run-out end of the pump operating envelope must be beyond the intersection with the head-capacity curve. This allows the pump to operate over the widest variable-speed range, so that at minimum speed the lead pump would discharge the minimum flow. This is illustrated in Figure 4.26.

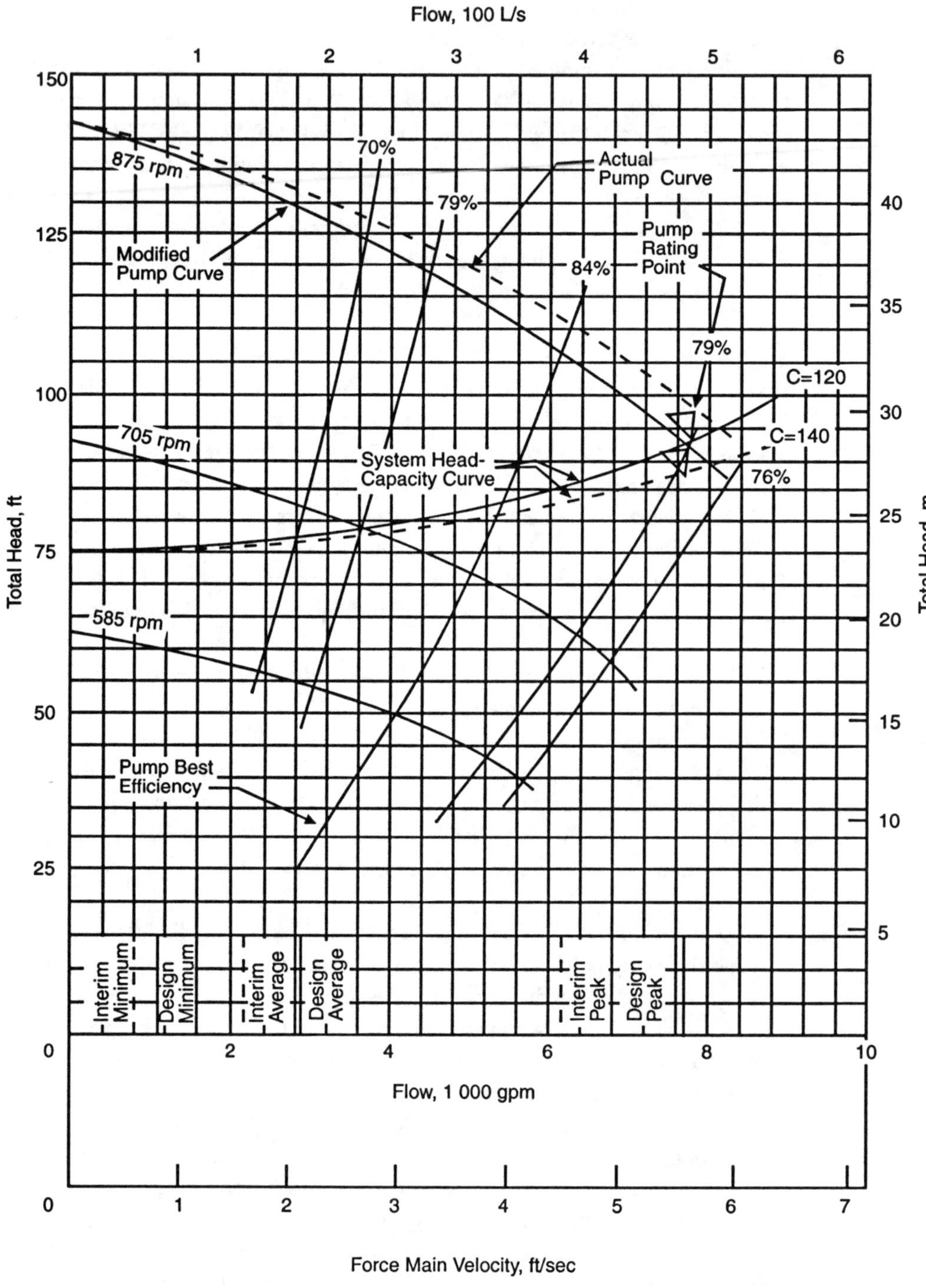

Figure 4.23 High static head system curve, one pump (ft/sec × 0.304 8 = m/s).

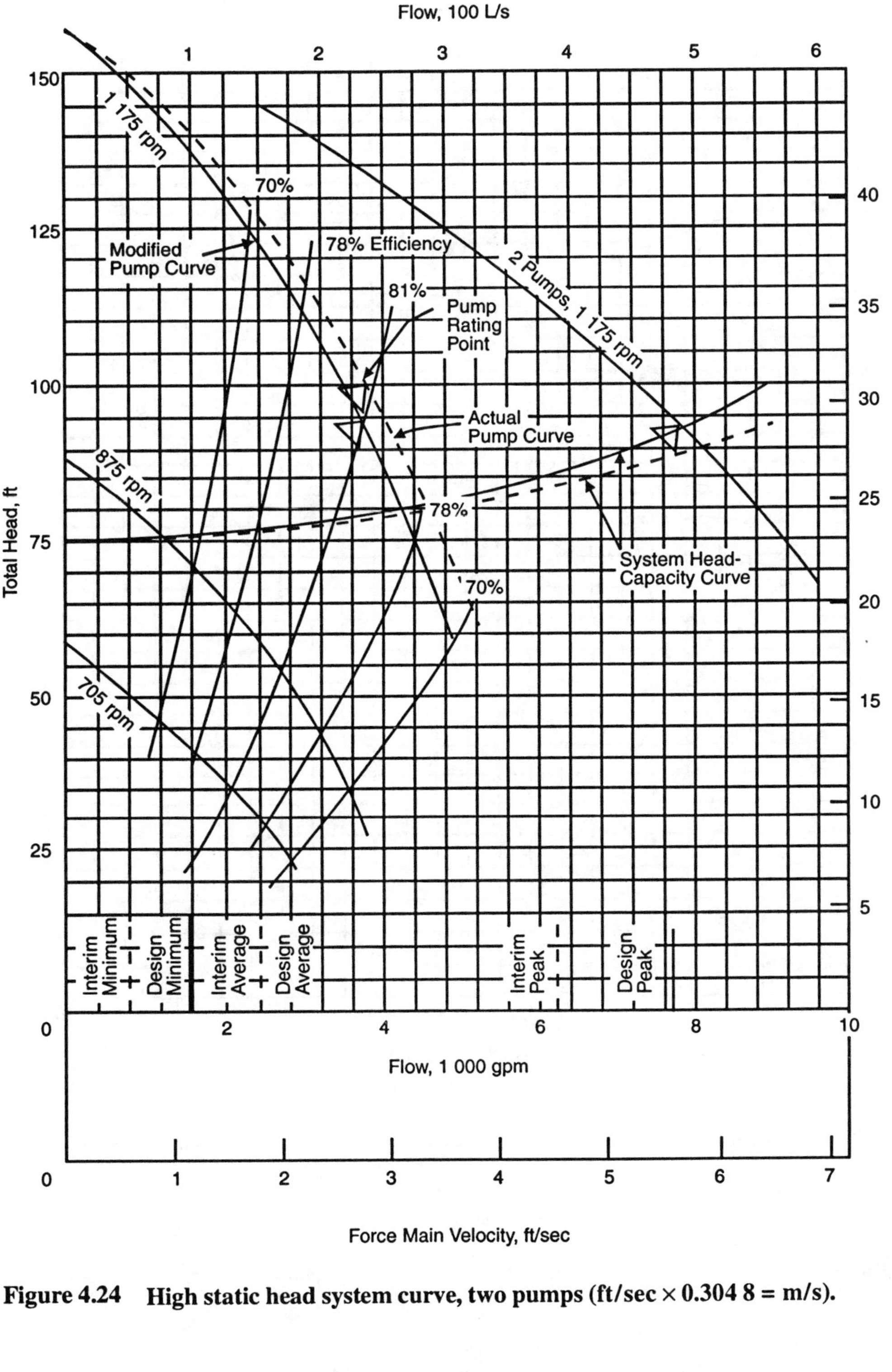

Figure 4.24 High static head system curve, two pumps (ft/sec × 0.304 8 = m/s).

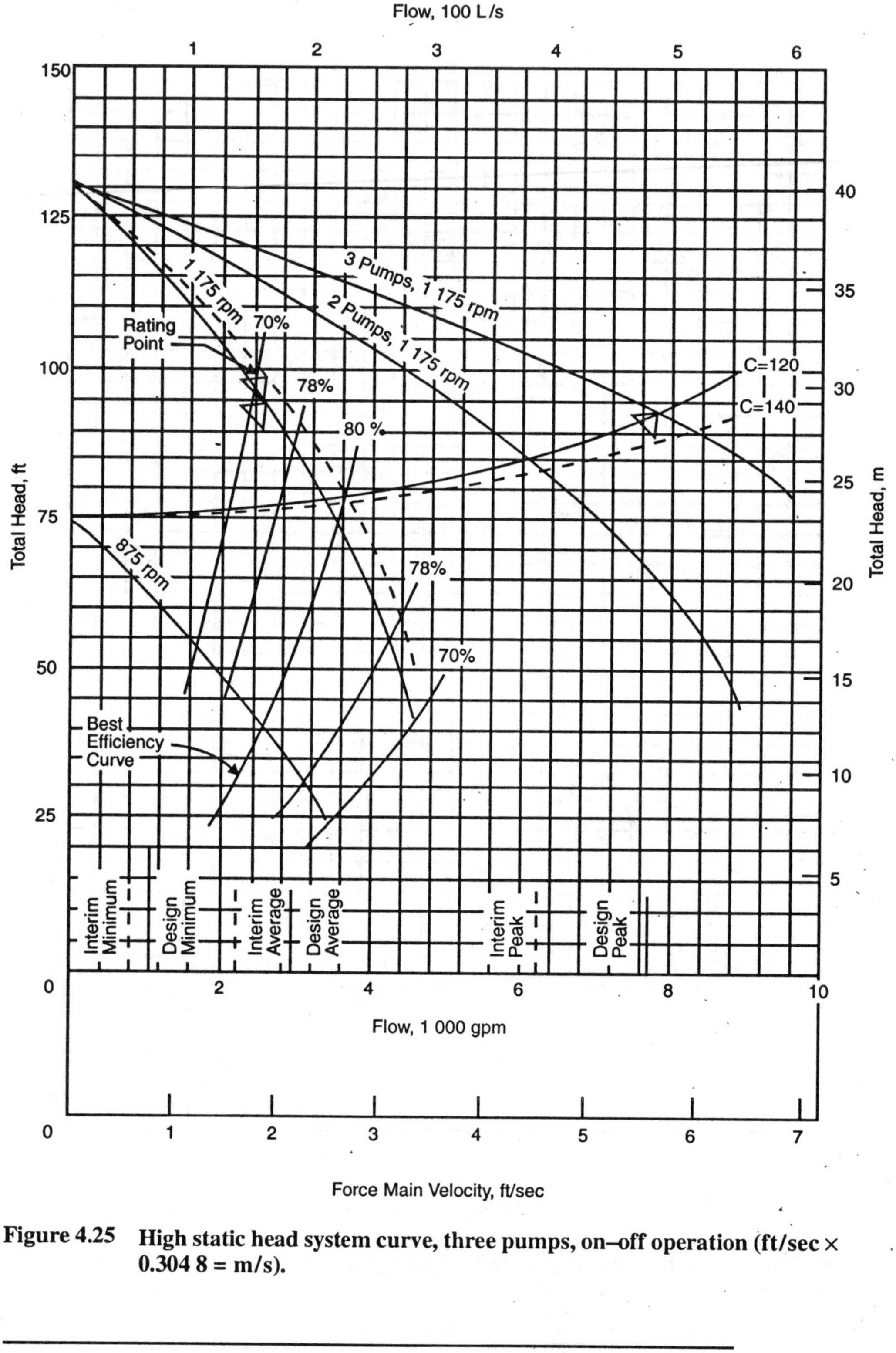

Figure 4.25 High static head system curve, three pumps, on–off operation (ft/sec × 0.304 8 = m/s).

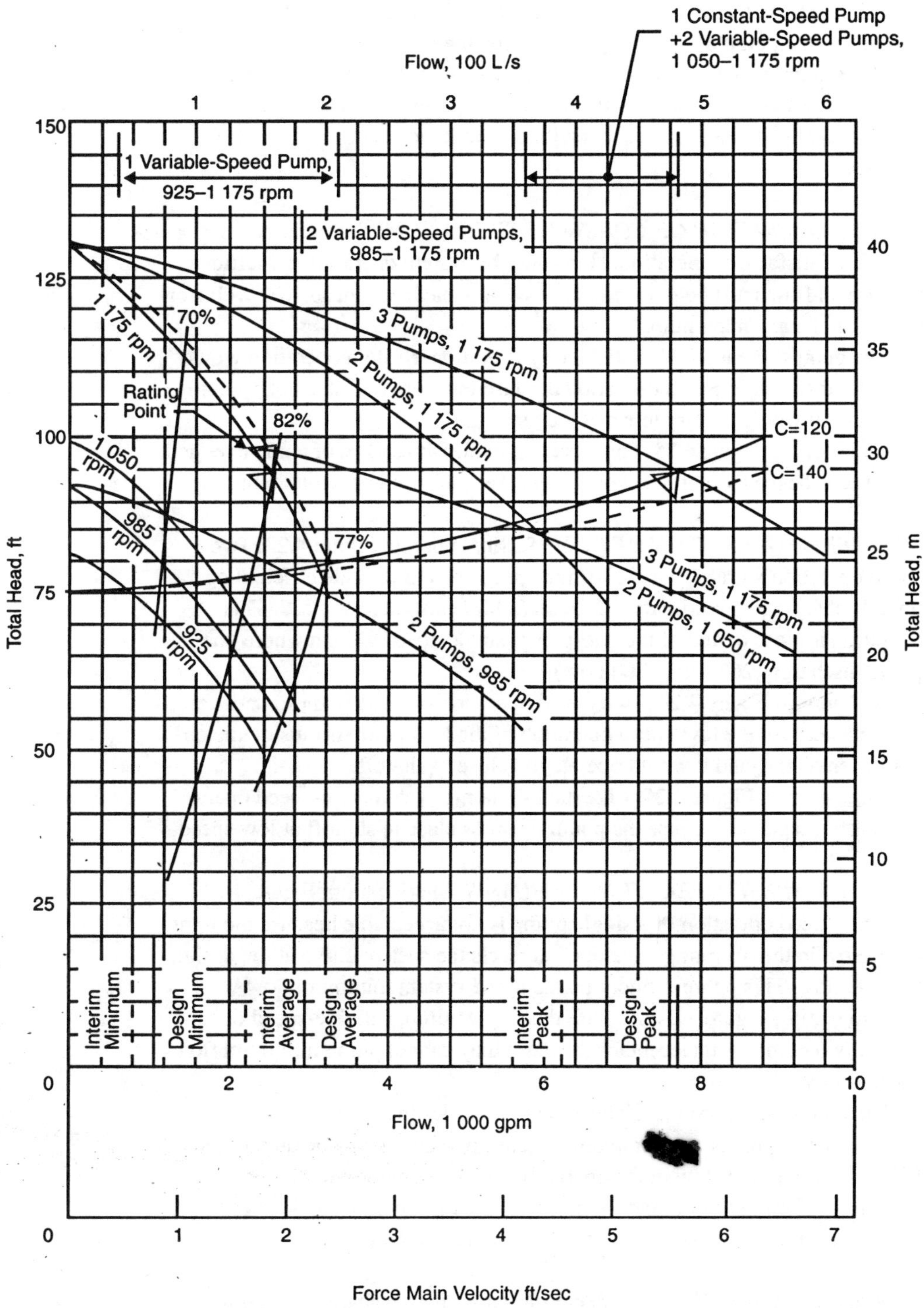

Figure 4.26 · High static head system curve, three pumps, variable-speed operation (ft/sec × 0.304 8 = m/s).

It should be noted on Figure 4.26 that when discharging the interim minimum flow, the pump operates slightly outside the pump operating envelope at 50 L/s (800 gpm). This may be acceptable because the pump would operate at reduced speed and head and at this flow for only short periods of time.

An analysis of these four curves with different modes of pump control follows.

ON–OFF OPERATION OF CONSTANT-SPEED PUMPS. Single pump, Figure 4.23. Satisfactory operation. However, the system operation is inefficient because all flow must be pumped at the system maximum head. Wet well storage would be at a maximum.

Two pumps, Figure 4.24. Satisfactory operation. System operation would be better than single-pump operation because there are two steps of pumping. This system would require less wet well storage.

Three pumps, Figure 4.25. This system provides three steps of pump rates and would provide the most efficient operation. Wet well size is at a minimum.

TWO-SPEED OPERATION OF PUMPS. Single pump, Figure 4.23. Two-speed operation with 875- to 705-r/min speeds would provide efficient operation, discharging approximately 250 L/s (3 800 gpm) at low speed. This operation is close to that of the two-pump, single-speed system where one pump discharges 265 L/s (4 200 gpm).

Two pumps, Figure 4.24. Two-speed operation of these pumps is not acceptable because the low-speed operation of the lead pump is outside the operating envelope, and it would operate too close to shutoff.

Three pumps, Figure 4.25. Like the two-pump system, two-speed operation is unacceptable because the pump operates close to shutoff at low speed.

VARIABLE-SPEED OPERATION OF PUMPS. Single pump, Figure 4.23. Variable-speed operation of a single pump is not acceptable because the minimum flow in the operating envelope intersects the system curve at the design average flow. This pump cannot operate to the system minimum flow.

Two pumps, Figure 4.24. As with the single pump, variable-speed operation of two pumps is unacceptable because they cannot pump minimum flows successfully.

Three pumps, Figure 4.26. This system provides the best operation for variable-speed pumps. At minimum system flow, the pump operates slightly below the minimum flow in the operating envelope, but it would operate at these flows for minimum time. In this case, the minimum flow dictates the size of the pump for a variable-capacity pumping system.

POWER CONSUMPTION. Power consumptions for the alternate systems are summarized in Table 4.5. The time required for pumping at the various flows is based on the same assumed diurnal flow pattern used for Figures 4.21 and 4.22.

Because all flows are pumped at maximum head, a maximum power draw of 65 000 W (1 550 kWh/d) is required by a single, constant-speed pump.

Power can be saved by the use of a two-speed motor for the single pump or by using two or three constant-speed pumps. These systems operate so that the majority of flow is pumped at less than maximum head and the power draw of the three alternates are approximately equal.

If variable-speed pumping is required, the power draw is slightly higher than on–off steps because of the power loss in the variable-frequency drives. The power draw in the three-pump system is less than the one- and two-pump systems because only one pump operates continually, with the second and third pumps operating for only short periods of time.

In summary, on–off control of a single pump is the simplest method of control but consumes the maximum power and requires the largest wet well storage volume.

Depending on the system curve, multiple pumps or two-speed operation of a single pump will reduce power consumption and required wet well storage, but the controls are more complicated than those of a single pump.

The use of variable-speed pumping to match inflow reduces required wet well storage, but has the most complicated control system and, depending on the system curve, does not reduce power consumption. Variable-speed pumping does, however, match inflow and when pumping to a treatment plant reduces shock loads on the treatment process.

Table 4.5 Daily power consumption—high static system.[a]

System operation	Single pump, kWh/d[b]	Two pumps, kWh/d	Three pumps, kWh/d
Single speed (on–off)	1 550[c]	1 360	1 350
Two-speed operation	1 350	d	d
Variable speed	1 650[e]	1 525[f]	1 450

[a] Power consumption based on an assumed diurnal flow variation for a system with a population of 36 000.

[b] kWh/d × 41.67 = W.

[c] The pump shown in Figure 4.22 draws 1 550 kWh. If a larger pump with the BEP at the rating point were used, the pump would draw 1 440 kWh.

[d] Two-speed pumps should not be used because low-speed operation requires pumps to operate at or near shutoff.

[e] Pump should be operated on–off at minimum speed for flows less than 165 L/s (2 600 gpm) to keep pump operation within the pump operating envelope.

[f] Pump should be operated on–off at minimum speed for flows less than 115 L/s (1 800 gpm) to keep pump operation within the pump operating envelope.

Large Pump Station System. Figures 4.27 and 4.28 present the pumping system for a large pump station where the major portion of the system head is static. This system curve is based on a new pump station for a large midwestern city that was built to activate a new relief interceptor.

The interceptor receives mostly sanitary wastewater; but, during wet-weather periods, large quantities of stormwater can be diverted to this interceptor. The contributing population is constant so there is only moderate increase in dry-weather flows between initial and design-year flows. There will be an increase in wet-weather flows as additional stormwater flows are diverted to the interceptor.

The pumping station is a lift station because it receives flow from the interceptor, lifts the flow, and discharges to an open channel. Therefore, the system head is constant, with its only variation being the difference between high and low wet well level for the speed control band.

System dry- and wet-weather flows are shown on Figure 4.27. To avoid confusion, only the modified pump curve is shown on this system curve.

The pumps are sized for seven pumps, including one standby, to discharge the design-year peak storm flow. Because the station discharges to a treatment plant, the pumping station has variable-speed pumps to match the influent flow.

The pumps are selected with the rated pump capacity beyond the pump BEP. Therefore, as the pump slows from full speed, the operating efficiency will increase until the best efficiency is reached and then decrease as the pump approaches the minimum flow in the pump operating envelope. The selected pump curve is shown on Figure 4.28.

Because the interim design wet-weather flow is only 17.2 m^3/s (395 mgd), five pumps will be installed initially, with space for two future pumps provided in the station.

Of the five pumps originally installed, four are furnished with variable-speed drives and one with a constant-speed motor. Future pumps will have constant-speed motors.

The reason for four variable-speed pumps is to allow operation without unnecessary wear to the pumps. The first two pumps on line must be variable speed to allow the pump rate to match the inflow without one pump operating near shutoff head. In this station, the minimum flow requires the operation of two pumps at low speed. However, these two pumps, operating at full speed, can pump more than the dry-weather design average flow. Therefore, if only two variable-speed pumps were provided, they would be operating continuously and the constant-speed pumps would operate only for short periods of time for flows greater than design average flows.

Three variable-speed pumps could be furnished, but each would operate for two-thirds of the time. Four variable-speed pumps are furnished to provide reliable station operation with one out of service. The variable-speed pumps not used for lead or first lag would be used as constant-speed pumps when required for peak flows after all the constant-speed pumps are in use.

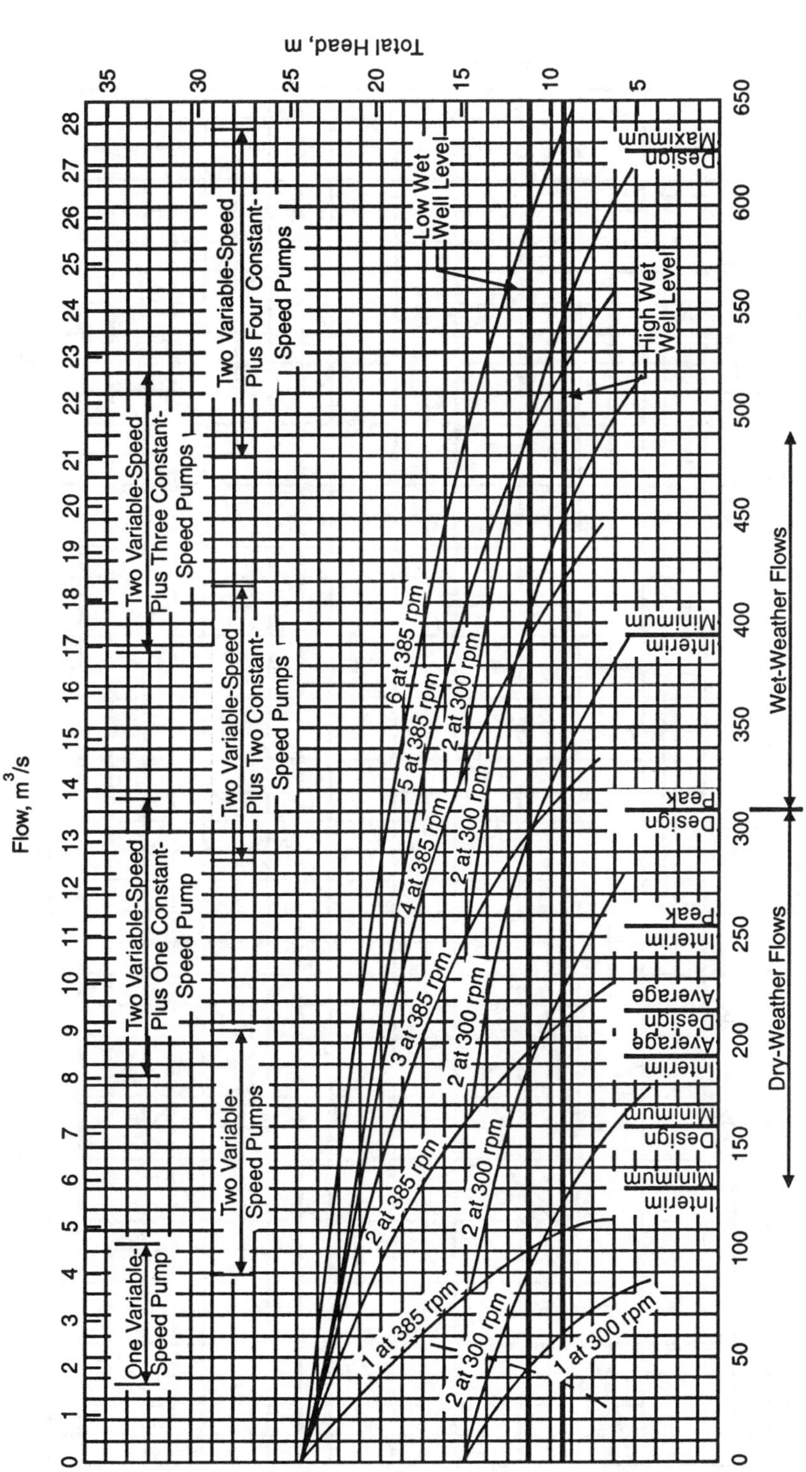

Figure 4.27 Large pumping station system curve.

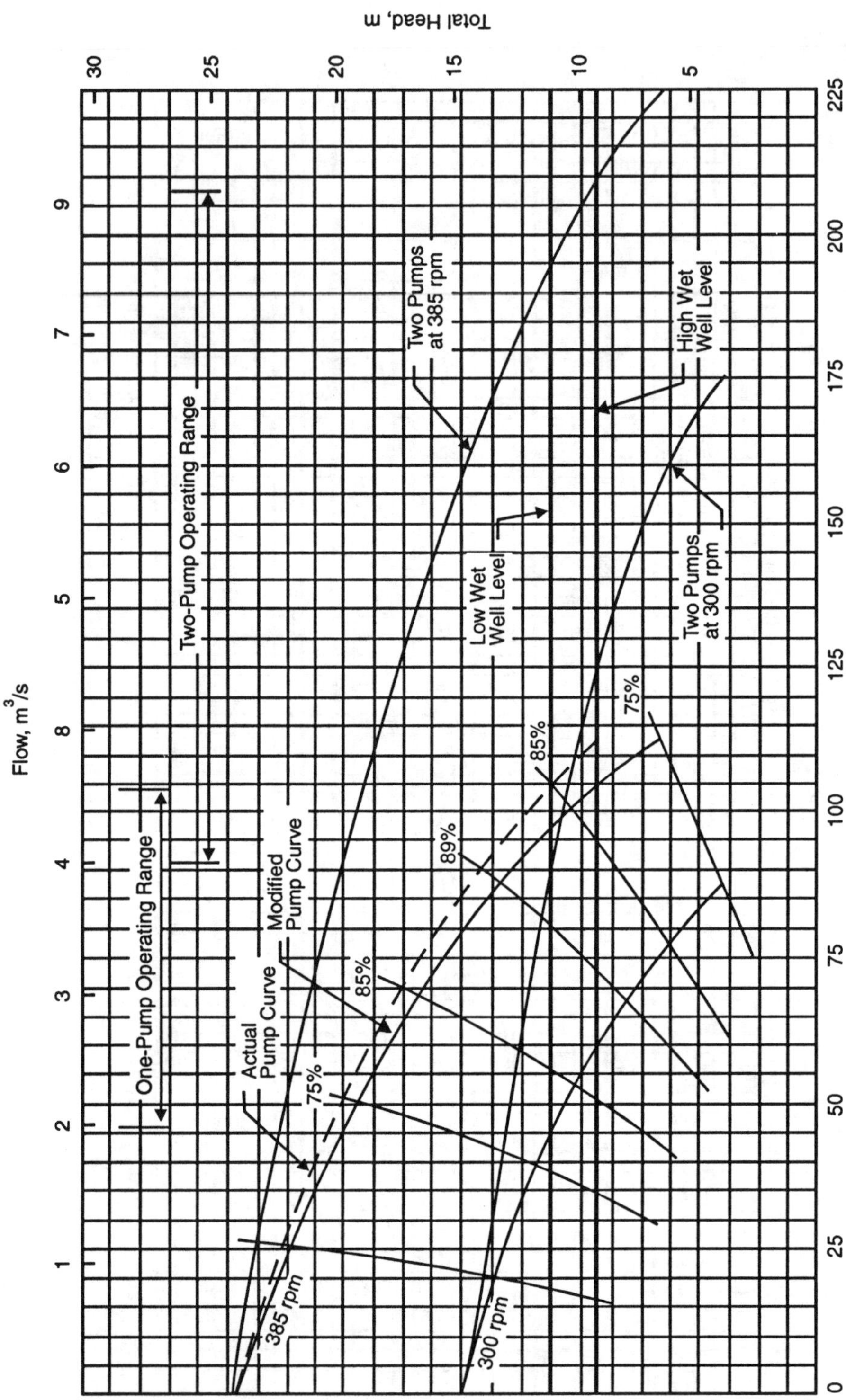

Figure 4.28 Large pump curve.

SPECIFYING PUMP PERFORMANCE

It is important to specify pump performance so that manufacturers have sufficient information to select pumps that meet the requirements of a pumping system. In addition to the rating point and maximum speed of a pump, the manufacturer must know its operating range and the system suction conditions.

All pumps can meet one rating point but not two. The rating point is typically the pump capacity required to discharge its portion of the design peak flow at the maximum system head, and pumps typically will operate at flows greater than the rating point. If the shape of the pump curve is critical, the engineer can specify alternate rating points that describe the allowable envelope but should not specify an exact point. An acceptable method of specifying an alternate rating point would be to specify "not less than ___ L/s (___ gpm) or more than ___L/s (___ gpm) at ___ m (___ ft) head."

To avoid cavitation, the manufacturer must also know the suction conditions. This information can be supplied by requiring successful operation to a specified minimum head or, in the case of a high-friction curve, successful operation to a curve between two alternate rating points.

The manufacturer must also know the NPSH available in the system. It is recommended that the $NPSH_A$ be understated by 1 or 2 feet. Hydraulic Institute Standards (1975) on testing allow a pump to exceed the rated flow by 10% at rated head or the head by 5% at rated flow. The additional friction losses at this plus tolerance may require the 1- or 2-ft safety factor in the $NPSH_A$.

In variable-speed pumping systems, the manufacturer must know the expected speed range, which can be provided by specifying a reduced speed rating point. All manufacturers can meet this second pump rating point, but the speed between manufacturers will vary.

In complex systems, the designer may consider including a rating curve in the pump specifications. This curve should include the pump rating point, alternate rating points, minimum and maximum system head curves, and system $NPSH_A$.

It is recommended that the specifications require all pumps be tested by the manufacturer. These tests should include a hydrostatic test and an operating test. The operating tests on small pumps typically are simple tests that prove the pump discharges the rated capacity at the rated head. Larger pumps should be tested throughout the required operating range of the pump with certified test curves submitted for review.

The need for witnesses to test the pumps should be considered by the designer, particularly for large pumps at critical installations. This requirement should be spelled out clearly in the specification as it would incur extra costs for the manufacturer, who typically tests these pumps twice—to prove pump operation before the witnessed test and during the actual witnessed test. Before the witnessed test, the manufacturer should submit test stand data, which must be reviewed to ensure that the test meets the requirements of Hydraulic Institute Test Codes and that the test stand can produce the required $NPSH_R$ to operate the pump over the specified range of operation.

REFERENCES

Design of Municipal Wastewater Treatment Plants (1992). Manual of Practice No. 8, Water Environ. Fed., Alexandria, Va.; ASCE Manuals and Rep. on Eng. Practice No. 76, Am. Soc. Civ. Eng., New York, N.Y.

Hydraulic Institute Standards for Centrifugal, Rotary and Reciprocating Pumps (1975). 14th Ed., Hydraul. Inst., Cleveland, Ohio.

Karassik, I.J., *et al.* (1976) *Pump Handbook.* 2nd Ed., McGraw–Hill, Inc., New York, N.Y.

SUGGESTED READINGS

Metcalf and Eddy, Inc. (1981) *Wastewater Engineering: Collection and Pumping of Wastewater*. G. Tchobanoglous (Ed.), McGraw–Hill, Inc., New York, N.Y.

Sanks, R.L., *et al.* (1989) *Pumping Station Design.* Butterworth Publishers, Stoneham, Mass.

Chapter 5
Piping Systems

The pumping station piping system discussion will follow the pool-to-pool concept, in which one analyzes the piping system between locations (pools) where the pumped liquid is affected by atmospheric pressure on the suction and discharge ends of the station being analyzed. There may be more than one pool on the discharge piping end as in multiple-pump stations or discharge points.

GENERAL DESIGN CRITERIA

The pumping station piping design must allow for wastewater or stormwater inflow variations. The flow will vary throughout the day, week, and season, and the station must be capable of handling peak flows with the largest pump out of service. The original design should consider future demands on the station, perhaps by installing connections for future pump suction lines. Discharge manifolds for larger pumps can be designed by providing for the replacement of pipe and by reducing fittings.

Pumping stations often serve far beyond their initial design life, and site selection should permit expansion, including wet well size and inlet piping for dry-pit stations. Sizing for the ultimate capacity of the station should be based on accepted standards (Recommended Standards, 1990) or governing regulations. To ensure the satisfactory operation of pumps without vortexing or air entrainment in the liquid, pipes entering the wet well should be constructed considering the ultimate capacity of the station.

PIPING. Piping is sized so that the velocity in the suction line will vary from 0.6 to 2.5 m/s (2 to 8 ft/sec), with a preferred value of 1.5 m/s (5 ft/sec), and will vary in the discharge piping between 1 and 2.5 m/s (3 and 8 ft/sec). As a general rule, when the velocity head ($V^2/2g$) exceeds 0.32 m (1 ft) for wastewater and 0.8 m (2.6 ft) for stormwater, a larger pipe should be used. Except for small grinder and effluent pump installations, piping less than 75 mm (3 in.) in diameter should not be used for conveying solids-laden wastewater.

VALVES. Vacuum or air-release valves should be considered for installation at high points in discharge piping. Sizing of these valves (air flow rate) should consider the surge effect on the system. Drain or blowoff valves at all low points would help in removing sedimentation.

Suction and discharge pipe should be supported and properly anchored and restrained. The use of expansion joints and flexible pipe couplings should be limited because they encourage poor piping practice. Flanged or grooved connections are recommended to facilitate dismantling and allow for thermal movements. Ductile iron flanged pipe or flanged fittings, if encased in concrete, will complicate repair work.

Both the suction and discharge piping must be supported and restrained to protect the pump casing. Settling of pipes may cause misalignment after installation, and care must be given to proper thrust blocking and anchorage of pipe where expansion or flexible joints are used and at pipe bends. Expansion or flexible joints must have piping restrained on both sides.

The capital cost of pipe and fittings in a large pumping station is a major part of the total cost; thus, the choice of pipe size and materials is important. Larger pipes may have higher initial capital cost, but smaller pipes lead to increased hydraulic losses and energy costs.

Suction pipe openings extending through the wall of the wet well should be cast with wall castings or sleeves and sealed to the concrete, and the suction pipe should be level or have a slight rise toward the pump to avoid air pockets.

Suction piping should be arranged so it can be cleaned through openings in the pump suction or piping. Each pump should have a separate suction line from the wet well fitted with a bell-type elbow fitting, properly located in relation to the wet well floor (Hydraulic Institute Standards, 1983). Horizontal lines without suction bells might suffice, depending on wet well design.

The gate or plug valve on the suction line should be of high quality to resist corrosion and still function when closed. The valve should be used regularly, and piping also should be designed to resist corrosion. Ductile iron pipe normally is supplied with cement lining, though coal tar coatings and plastic or epoxy linings also can be used. Reducers on the suction side of pumps should be eccentric and installed with the top flat to prevent air entrapment. A sluice gate installed at the wet well entrance and between sides of a divided wet well facilitates dewatering for cleaning, inspection, and repair.

MAINTENANCE. To allow for proper maintenance of the pump and check valve (or pump control valve), a gate or plug valve for isolation should be provided in the discharge line of each pump downstream of the check valve. Check valves should be installed in a horizontal line as opposed to a vertical line.

FORCE MAINS. Though force mains generally are designed to reduce annual operating costs, the designer also should consider the possible accumulation of deposited material in the line. Optimum velocities for reducing cost and preventing accumulation of solids range between 0.9 and 1.5 m/s (3 and 5 ft/sec), which are high enough to resuspend material that may settle out when the pumps are not operating. The use of insertable mechanical cleaning devices called "pigs" to clean the discharge lines should be investigated in all new installations.

For years, liquid pipelines have been designed for periodic and automatic cleaning using the "pigging" procedure, in which a bullet-shaped projectile of roughly the same diameter as the pipe inside diameter is launched through the pipe from a special entry station and cleans as it moves slowly through the length of the pipe to a recovery station. For longer pipelines, the cost effectiveness of the concept should be investigated. The cost-benefit analysis is essentially a constant friction loss and power cost versus an increasing power cost with increased friction losses. Increased pipe roughness is a highly selective factor differing for each design and transported liquid.

The increasing popularity of variable-speed pumping must be considered in relation to force main velocity, and provisions for cleaning or flushing must be made for extended low flow time periods with velocities less than 0.6 m/s (2 ft/sec). One suggested method is manual operation at higher speeds and at a predetermined time interval.

Materials used for force mains include ductile iron, steel, polyethylene, polyvinyl chloride (PVC), fiber glass or reinforced plastics, and prestressed and reinforced concrete. System design pressures and surge allowance may preclude the use of some materials.

Flow typically is intermittent in wastewater force mains, with gases (predominantly H_2S) generated during periods of nonflow. When combined with water, a mild sulfuric acid is formed that is corrosive to many pipe materials.

In longer force mains, especially at higher ambient temperatures, wastewater generates hydrogen sulfide gas (H_2S) from anaerobic decomposition. Controlled injection of air or chemicals to the force main at the pumps in larger pumping stations can reduce hydrogen sulfide buildup. Care must be taken, however, in selecting the point of entry and the rate of injection. Some alternatives to air injection or chemical feed are hydrogen peroxide and ozone or the use of noncorrosive force main piping materials (WPCF, 1979). Installation of an isolation valve on the force main just outside the pumping station would allow the valve installed on the pump discharge to be inspected and serviced, assuming the discharge line can be taken out of service. A bypass connection outside the pumping station would permit bypassing the station with a portable pump.

Good practice dictates pressure testing portions of force mains, which are completed in advance of the entire system. Before testing, the pipeline must be backfilled or braced to prevent movement under pressure. Pipeline bends must be fitted with permanent thrust blocks or harnessed restrained joints, constructed to withstand test pressure thrusts. When testing a partially completed pipeline, hydrostatic testing is preferred over air testing for safety reasons.

The force main should be filled slowly from any available source, with water introduced from temporary connections made at the lowest point in the line. Flow velocity during line filling should not exceed 0.6 m/s (2 ft/sec), depending on the pressure rating of the system.

Air and vacuum valves should be designed for wastewater service with cleanout provisions to facilitate maintenance, as they are likely to clog with grease from force mains carrying raw wastewater. On low spots where gritty material is likely to accumulate and restrict flow through the force main, a blowoff should be installed. Points of discharge should be well ventilated. To reduce corrosion, force mains should be kept filled during periods of normal intermittent operation and dewatered only for maintenance or repair. Air vent and relief valves should be protected from freezing and from damage by heavy equipment such as is used for construction and grass cutting.

FLUID FLOW FUNDAMENTALS

Pumping stations are used where it is not possible or cost effective to construct a gravity line to remove stormwater or wastewater. Often called lift stations, pumping stations raise wastewater from one gravity line to another. Figure 5.1 illustrates the schematic factors involved in assessing the hydraulics of a stormwater or wastewater pumping station. Typically, the liquid is to be transported from one location to another at a higher elevation. A pump is needed to produce the required total head at the required flow rate. The total head is the sum of the pipe friction losses and the elevation difference from the inlet water level to the highest elevation in the discharge force main or static head.

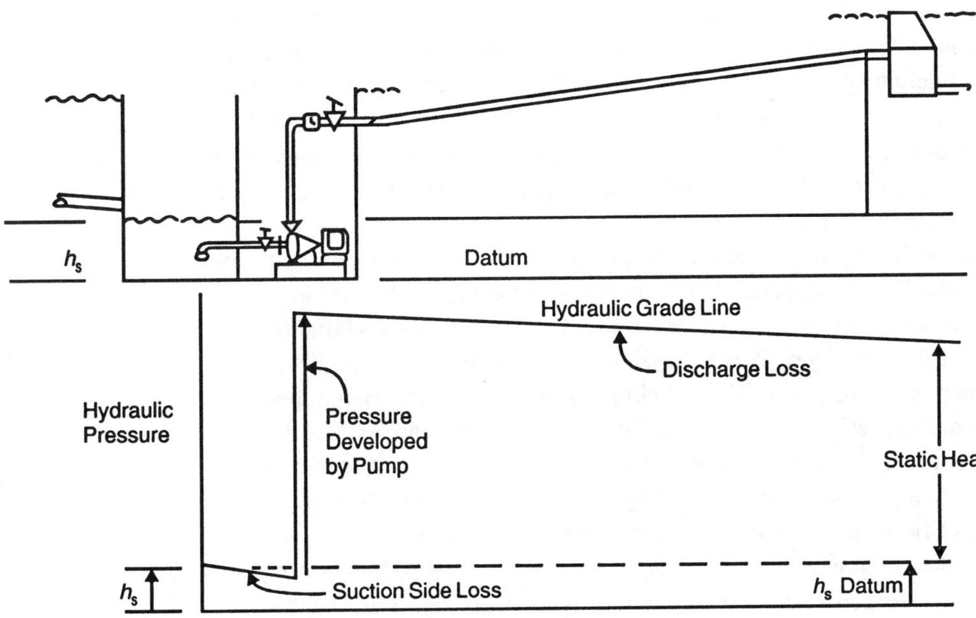

Figure 5.1 Vertical elevation system hydraulic grade line: h_s = suction head.

The energy level of the liquid from an elevation datum and losses in the system are shown on the hydraulic gradeline in Figure 5.1. The energy grade line reflects a certain system condition such as a single flow rate. When designing a pumping station, a designer should establish a range of flows and select a specific pipe size and station arrangement. The resistance to flow is represented by the head-flow curve resulting from this analysis and is called a system resistance or system head curve.

SYSTEM CURVE. The system head or resistance curve is a picture of the total head (pressure) a pump must supply when operating to produce a certain flow rate. The total head the pump must develop is the sum of the static head and friction losses. The pump can only operate at the intersection of the system head curve and pump performance curve.

Pipe friction losses can be determined by the selection of roughness coefficients such as Hazen–Williams, Darcy–Weisbach, or Manning, which then are used in the appropriate equations. When determining economics, roughness coefficients typically are based on pipe roughness after use for several years. Pump selection, however, should consider difference in suction pressure and the pipe roughness for both new and old pipe. The friction head should be calculated as accurately as possible. Fitting or minor losses are calculated using applicable coefficients to apply to the velocity head for the type of fitting or valve being considered.

The economic diameter of a force main is selected based on the size that will give the lowest overall annual cost or capitalized cost, depending on the method of calculation. The procedure involves selecting various pipeline diameters and calculating the pump head based on the velocity of pumping at the normal rate of flow (not the peak rate). The pump is sized from its head, and the pumping power is determined from the pump curve at its corresponding head and efficiency for the normal flow rate. The pumping power cost then is calculated using both the energy per kilowatt hour and the demand charge per kilowatt. This cost may be capitalized and added to the capital cost for the pump, piping, and force main.

Alternately, the capitalized cost of the latter three items may be calculated on an annual basis and added to the annual power cost. The pipe diameter that gives the lowest cost, on either basis, is the most economical diameter providing the velocity of flow is suitable for wastewater pumping. Several computer programs are available to simplify pipe size selection, pump selection, and power calculations (Krebs, 1990; Micro, 1986; and Wood, 1992).

Pressure surges are any deviation from the normal steady-state hydrostatic pressure in a pumping system, and they occur during partial or complete interruption of flow for reasons including pump starting and stopping, valve closures, power failures, valve openings, or if pipes are broken. Most stormwater and wastewater force mains are under low hydrostatic pressures, but the flows have relatively high operating velocities and momentum that contribute to high surge pressures.

In the design of pipe, care should be taken to determine wall thickness based on internal operating pressure, surge pressure, and total (passive and active) earth or equipment loads on the pipe. Also, proper bedding of the pipe must be provided. Joint design for each type of pipe should be investigated for proper compression of gaskets, expansion and contraction of joints, and the thrust anchors. The weak links in many force mains are at bends where thrust restraints are needed, and the maximum pressure the pipeline will be subjected to should be determined carefully.

EQUATIONS. Liquid flow relationships in piping systems are described by three mathematical relationships:

- The continuity equation;
- The general energy equation (Bernoulli); and
- The momentum equation.

While these relationships are based on the assumptions of incompressibility of the pumped fluid and steady-state or one-dimensional (stream tube) flow, equations are commonly applied to all conditions of pipe flow. Surge analysis requires special consideration. Under steady-state (constant at any one point in time) flow conditions, the continuity equation states that the product of flow area and velocity is constant for a given flow rate. The general energy equation is credited to Bernoulli and applies to the energy level of a fluid at any particular location in a system. If the liquid is incompressible and no heat is added or taken away and no work is done by or on the liquid, then the appropriate energy terms are velocity head, pressure head, and position or elevation head.

In real systems, between two points there would be an energy change so that comparing the terms between two points will give a differential energy term, head loss.

Finally, the momentum equation relates the change in velocity vectors (direction or magnitude change) to a mass rate of flow and the resulting components of force that must be restrained. A pipe bend is an example.

The continuity equation is as follows:

$$Q = AV \tag{1}$$

(For liquids)

$$Q_1 = Q_2 \tag{2}$$

then

$$A_1 V_1 = A_2 V_2 \tag{3}$$

Where

Q	=	flow rate, m^3/s (cu ft/sec);
A	=	area, m^2 (sq ft); and
V	=	velocity, m/s (ft/sec).

The general energy equation is as follows:

$$Z_1 + \frac{V_1^2}{2g} + \frac{P_1}{Y} = Z_2 + \frac{V_2^2}{2g} + \frac{P_2}{Y} + h_L \tag{4}$$

Where

h_L	=	energy loss or gain, m (ft);
Z	=	elevation above datum, m (ft);
$V^2/2g$	=	velocity head, m (ft);
g	=	gravitational constant, 9.8 m/s^2 (32.2 ft/sec^2);
P	=	liquid absolute pressure, kN/m^2 (lb/sq ft); and
Y	=	specific weight, kg/m^3 (lb/cu ft).

PIPE FLOW PARAMETERS. To properly size pumps for the station, the designer needs to calculate the pressure required to move the liquid through the pipe at the desired flow rates. Various investigators have presented equations for calculating friction losses in closed conduits. The resistance to flow in a pipe of any incompressible liquid is proportional to pipe size and velocity head. The friction factor is the constant of proportionality and is itself a function of pipe roughness.

The Darcy–Weisbach and Hazen–Williams formulas are used for filled pressure pipes, and the Manning equation is used for open channel and gravity flow in pipelines with a free water surface. Table 5.1 lists these equations in a form for open channel and full pipe flows. The formulas have been incorporated to tables for pipe size and type in many hydraulic handbooks (Cameron, 1992; Submersible, 1986; Hydraulic Institute Engineering, 1979; and Flow, 1974) and also have been included in many computer programs and for simplified pipe-sizing procedures (Wood, 1991; Micro, 1986; and KY Pipes).

Most tables of friction losses are calculated on the basis of new pipe. The designer must estimate and incorporate to the design the increase in pipe roughness from aging. The Hazen–Williams equation C factor (or aging coefficient) is a ready way to estimate increased pipe roughness for use in that formula. Table 5.2 shows C values typically used for design purposes. Note that new pipe values vary from the design C values.

The absolute roughness parameter varies with pipe type and surface finish. Table 5.3 provides the values of the roughness coefficient for different pipe types for use in the Darcy–Weisbach, Manning, and Hazen–Williams equations. The relationship between pipe roughness and friction factor can be obtained from the Moody diagram, Figure 5.2.

Table 5.1 Pipe friction loss equations.

Darcy–Weisbach:

$h = f[L/D]V^2/2g$

h	=	friction loss, m (ft)
f	=	friction factor
L	=	pipe length, m (ft)
D	=	pipe diameter, m (ft)
V	=	velocity, m/s (ft/sec)
g	=	gravitational acceleration 9.81 m/s² (32.17 ft/s²)

Hazen–Williams:

$h = L[V/0.85C \cdot R^{0.63}]^{1.85}$ metric

$h = L[V/1.318C \cdot R^{0.63}]^{1.85}$ English units

C	=	pipe roughness coefficient
R	=	hydraulic radius, m (ft) ($D/4$ for circular pipes)

Manning (open channel):

$h = L[Vn/R^{0.67}]^2$ metric

$h = L[Vn/1.49R^{0.67}]^2$ English units

n	=	roughness coefficient
R	=	hydraulic radius (cross-sectional area wetted perimeter, m (ft)

Table 5.2 Hazen–Williams friction coefficients (*C* for condition of pipe).

	C		
Pipe type	**New**	**Design**	**17–20 years**[a]
Ductile iron (lined)	150	140	120
Fabricated steel	140	130	100
Fabricated steel (lined)	145	135	120
Plastic—polyvinyl chloride	150	140	140
Concrete cylinder pipe	145	140	120

[a] Values can vary widely because of site-specific conditions such as velocity, length of line, and liquid characteristics.

Table 5.3 Values of effective absolute roughness and friction formula coefficients.

Conduit material	Effective absolute roughness (Darcy–Weisbach) k, ft	Manning n, $ft^{0.5}$	Hazen–Williams C
Closed conduits			
Asbestos-cement pipe	0.001–0.01	0.011–0.015	100–140
Brick	0.005–0.02	0.013–0.017	—
Cast iron pipe			
Uncoated (new)	0.000 85	—	—
Asphalt dipped (new)	0.000 4	—	—
Cement-lined & seal-coated	0.001–0.01	0.011–0.015	100–140
Concrete (monolithic)			
Smooth forms	0.001–0.005	0.012–0.014	—
Rough forms	0.005–0.02	0.015–0.017	—
Concrete pipe	0.001–0.01	0.011–0.015	100–140
Corrugated-metal pipe (0.5-in. × 2.6-in. corrugations)[a]			
Plain	0.1 –0.2	0.022–0.026	—
Paved invert	0.03 –0.1	0.018–0.022	—
Spun asphalt-lined	0.001–0.01	0.011–0.015	100–140
Plastic pipe (smooth)	0.01	0.011–0.015	100–140
Vitrified clay			
Pipes	0.001–0.01	0.011–0.015	100–140
Liner plates	0.005–0.01	0.013–0.017	—
Open channels			
Lined channels			
a. Asphalt	—	0.013–0.017	—
b. Brick	—	0.012–0.018	—
c. Concrete	0.001–0.03	0.011–0.020	—
d. Rubble or riprap	0.02	0.020–0.035	—
e. Vegetal	—	0.030–0.40	—
Excavated or dredged			
Earth, straight, uniform	0.01	0.020–0.030	—
Earth, winding, fairly uniform	—	0.025–0.040	—
Rock	—	0.030–0.045	—
Unmaintained	—	0.050–0.14	—
Natural channels (minor streams, top width at flood stage < 100 ft)	0.1 –3.0	—	—
Fairly regular section	—	0.03 –0.07	—
Irregular section with pools	—	0.04 –0.10	—

[a] in. × 25.4 = mm.

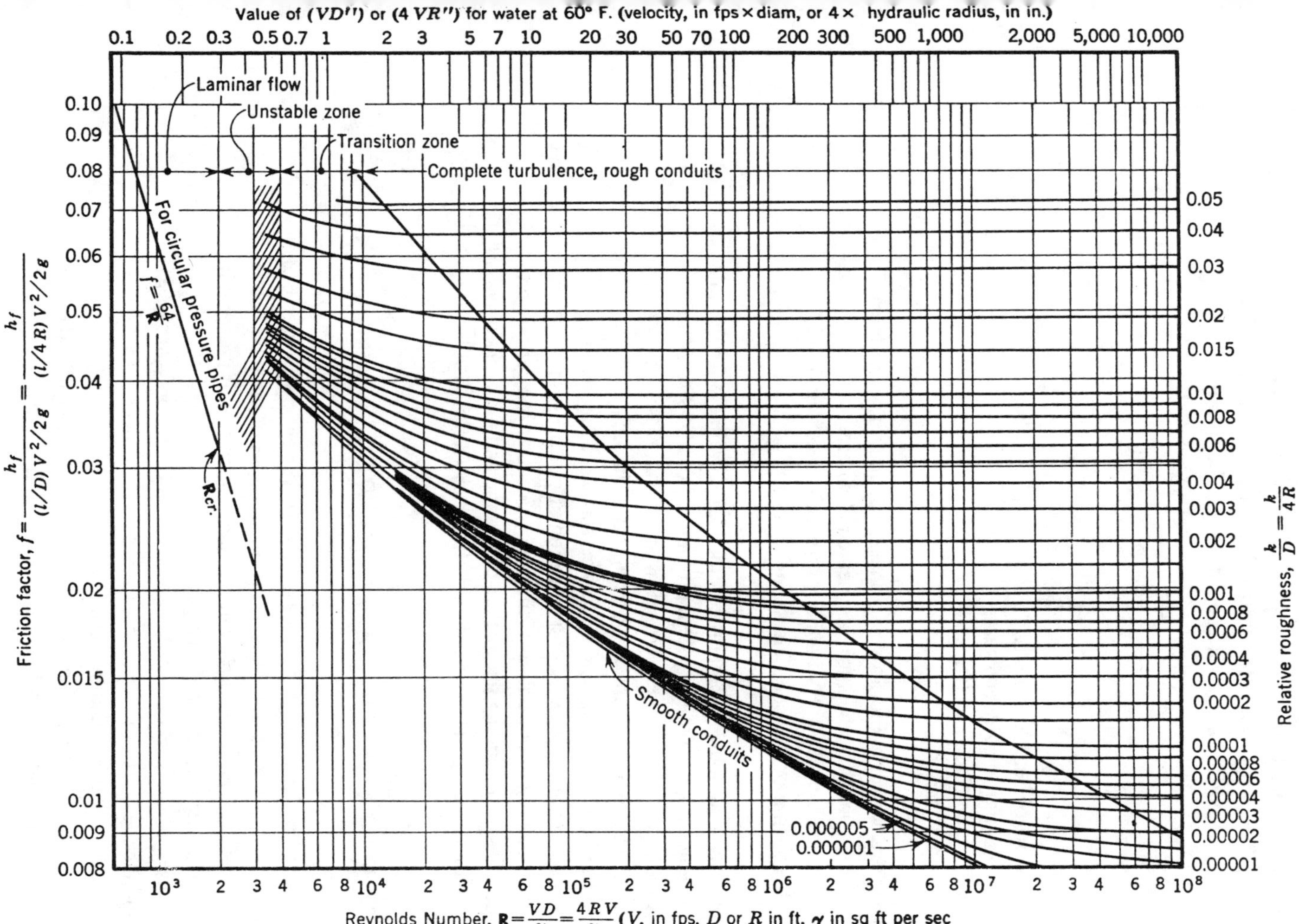

Figure 5.2 Moody resistance diagram for uniform flow in pipes [0.555 6(°F–32) = °C; in. × 25.4 = mm; ft × 0.304 8 = m; ft/sec × 0.304 8 = m/s; sq ft/sec × (9.29 × 10^{-2}) = m^2/s].

PIPE FITTING PARAMETERS. In most systems, frictional losses from pipe fittings normally represent only 10% or less of the total system friction losses. Though small, they must be calculated and added to the system. In stormwater pumping stations, minor losses can be a high percentage of total friction losses. Minor head losses (fitting losses) are proportional to the velocity head ($V^2/2g$). The proportionality constant K has been experimentally and analytically determined for a large variety of fittings. The formula is

$$h = \frac{KV^2}{2g} \tag{5}$$

Where

h	=	minor head loss, m (ft);
K	=	proportionality constant;
V	=	pipe velocity, m (ft); and
g	=	gravitational constant.

Some designers convert minor losses to an equivalent number of feet of straight pipe. There are also tables for this method (Cameron, 1992). Hydraulic handbooks (Cameron, 1992; Hydraulic Institute Engineering, 1983; and Flow, 1974) list the minor loss coefficients for common and more complicated shapes.

There are several computer programs commercially available for the solution of simple and complex piping system problems (Wood, 1991; Micro, 1986; and KY Pipes). Some of these programs are useful, reasonably priced, and can be used on personal computers. They greatly simplify the task of pipe sizing, pump head determination, and system performance prediction.

There are also computer programs for selection of centrifugal pumps, and some use pump manufacturer catalogs (Pump-Flo). Another is a generic program that specifies the operating parameters of a pump and produces a technically correct pump size, operating speed, and efficiency (Krebs, 1990).

WATER HAMMER AND SURGE

Rapid changes in flow can lead to pipe pressure upsurges or downsurges that can cause rupture or water column separation and collapse of the pipe. Water hammer is the name given to the rapid fall and subsequent rise of pressure in a pipeline that can result from rapid changes in flow.

The analysis of a piping system to determine if significant pressure surges can occur and the methods of mitigating or avoiding these destructive forces is a complex subject best handled by experts.

This chapter is a general but elementary discussion of the rigid and elastic water column theories and also includes a review of some of the system conditions that could cause the designer to need an expert in the field. References covering the subject include Hydraulic Institute Standards (1983), Parmakian (1963), and Surge—A Computer Analysis of Transient Flow.

Liquid flowing in a closed conduit or pipe has an energy level from the momentum of the mass flow. If the flow in the pipe is interrupted, the existing momentum of the flowing liquid will tend to keep the liquid in motion until dissipated. The momentum energy is converted to pressure energy waves that travel back and forth through the length of the pipe until dissipated. Pressure changes caused by an interruption or other change in flow will be transmitted through the pipe at the speed of sound or celerity (sonic velocity) for the liquid and pipe combination. Typical values of celerity or speed of sound of water in pipes range from 305 to 1 220 m/s (1 000 to 4 000 ft/sec), with values increasing as material rigidity increases.

All pump piping systems subjected to changes in flow will generate some water-hammer-type pressure surges, though some may not be severe enough to cause failure or long-term damage to the system.

CHECKLIST FOR SURGE PROBLEMS. There are numerous events and conditions associated with pipeline system design of a pump station that indicate a need for a surge analysis, including

- Electric power outages to pump motors or controls;
- Standby power systems cannot respond to power outages fast enough to mitigate pressure surges from a pump trip;
- The length of the pipeline gives a cycle period for pressure surges that is greater than a reasonable flow interruption time;
- Reverse flow through pumps;
- Forward flow through pumps after pump shutdown, such as in low head systems;
- System pressure in the pipeline is high in relation to pipe design strength before surge consideration;
- System pressures in the pipeline are low and vapor cavity formation is possible because of a significant elevation change, 10 m (33 ft) or more over a short distance;
- There are dead ends in the piping system conducive to pressure amplification through resonance;
- Frequent or significant elevation changes of the pipeline, 10 m (33 ft) or more over a short reach of pipe;
- High pipeline velocities subject to rapid fluctuations;
- The pumping system is being changed to much larger flows in an existing piping system;
- Rapid release of trapped air generates a pressure upsurge;
- A pipeline rupture for any reason, causing water hammer;

- Control valve and check valve closing and opening times; and
- Repetitive stress from pressure surges, resulting in fatigue failure of the pipe regardless of pipe material.

RIGID WATER COLUMN SURGE THEORY. Among the assumptions accompanying the rigid water column theory are that

- The pipe does not expand with the pressure waves;
- The liquid is incompressible;
- The pressure in the pipe remains above the liquid vapor pressure; and
- Velocity at all sections of the pipe is the same when flow interruptions occur.

If a pipe of constant area connects a reservoir and a distribution point, a valve near the distribution point opens to let liquid flow from the reservoir to the distribution point. When the flow is interrupted by closing the valve, the momentum of the liquid in the line will create an unbalanced force on the closed valve manifesting itself as pressure waves back and forth through the pipe until dissipated. The velocity of the pressure waves will be equal to the velocity of sound for the liquid and pipe.

The maximum pressure developed in this case can be shown to be equal to

$$H = V_c\, V/g \tag{6}$$

Where

H	=	maximum pressure head, m (ft);
V_c	=	sonic velocity, m/s (ft/sec);
V	=	the line velocity at time of interruption of flow, m/sec (ft/sec); and
g	=	gravitational constant, 9.8 m/s^2 (32.2 ft/sec^2).

This is known as the Joukousky relation.

To find the total pressure in the line, the static pressure must be added to the surge pressure.

The maximum pressure emerges when the flow is interrupted in a time interval of less than the cycle period. The pressure wave is traveling at sonic velocity, and the period of time for the wave to complete its cycle to one end of the pipe and return can be shown to be

$$T = 2L/V_c \tag{7}$$

Where

L	=	pipe length, m (ft), and
T	=	cycle period in seconds.

PRESSURE WAVE VELOCITY

The velocity of the moving pressure waves will equal the velocity of sound for the liquid–pipe materials combination. For the liquid–pipe medium, a convenient approximate formula is

$$V_c = 4\,660/[1 + CR]^{0.5} \tag{8}$$

Where

C	=	bulk modulus of liquid/bulk modulus of pipe material and
R	=	pipe inner diameter/pipe thickness.

The pressure wave velocity varies somewhat with the anchorage of the pipe against longitudinal movement (Parmakian, 1963).

An approximate modulus of elasticity for several pipe materials at standard conditions and a value of the celerity of water in pipes without axial restraint is listed in Table 5.4.

ESTIMATING MAXIMUM SURGE PRESSURES. By combining a velocity calculation and the Joukousky relation (equation 6), the maximum surge pressure can be estimated. The surge pressure thus determined plus the normal pipe pressure will equal the maximum pressure in the line.

Table 5.4 Modulus of elasticity and celerity in pipes without axial constraint.[a]

Pipe material	Modulus of elasticity, E, $MN/m^2 \times 10^{-3}$	Celerity, V_c, m/s
Steel	205	1 260
Iron	130	1 065
Concrete pressure pipe	—	1 065
Aluminum	65	945
Copper	110	1 100
Polyvinyl chloride, plastic	3	305
Fiber-glass-reinforced pipe, plastic	6	425
Water (only no pipe) (bulk modulus)	—	1 440

[a] V_c = sonic velocity; $MN/m^2 \times 145$ = psi; and m/s × 3.281 = ft/sec.

PROBLEM EXAMPLES. As noted in the checklist for surge problems, there are many conditions that can cause excessive transient pressures to develop in a system. Two common examples involve flow from a reservoir, which does not involve pumping, and the case of a pumping station with discharge line.

Where water flows by gravity from a reservoir, interrupted flow from a valve closing generates a pressure wave. This problem is solved by closing the valve over a time period greater than the cycle period of the pressure wave. Even though the valve is closed in a time longer than the cycle period, a reduced pressure surge may occur.

The closing rate of the valve must be field adjustable. Because a power outage may render the valve control ineffective, surge relief valves often are used to solve this problem. The surge relief valve must be sized to pass a determined percentage of the flow that depends on line velocity and pressure. Surge relief valves may not open fast enough to protect the system from the fast moving pressure wave.

Surge anticipation valves also can be used to overcome this problem. In some cases, two surge relief valves set to open at different pressures have been used.

In the case of a pumping station and discharge line, potential surge problems would be aggravated by rapid closing of the check valve if the pump stopped. The resulting pressure wave downsurge could open a check valve not resistant to such action, which could be controlled by springs, weights, or a controlled closing check valve.

The closing rate of the check valve is field adjusted so that the closing time exceeds the cycle period of the system or a greater time to eliminate surge. As an alternative, an isolation plug or ball valve can be equipped with an actuator. To allow maintenance of isolation valves, they should be in series with a guard valve (gate or plug).

The pump normal start and stop procedure is important. On call, the pump starts. When the pressure upstream from the check or isolation valve exceeds the system pressure, the valve opens at a controlled rate. On pump stop, a signal will start closing the check or isolation valve at a predetermined rate. When approximately 90% closed, the pump is shut down. The time to close the valve is related to the cycle period time for the system and must be field adjustable.

A pressure-relief valve or valves is the normal solution to attenuate high surge pressures generated by a power outage. The pump suction line can be the source of surges under some conditions, such as a longer suction line or high suction line velocity.

SURGE-ATTENUATING DEVICES

The equipment and methods used to reduce surge pressure waves are illustrated in Figure 5.3 and include

- Controlled closing check or pump control valve;
- Air exhaust and air inlet valves in the discharge line;
- Surge or pressure-relief valve;
- Pressurized surge tank;
- Surge towers;
- Bypass check valve; and
- Using the inertia of pump and motor drive with the possible addition of flywheel mass.

CONTROLLED CLOSING CHECK OR PUMP CONTROL VALVE. A positive method of avoiding pressure surges on pump start-up and shutdown is to control the rate of opening and closing of the discharge check valve or, alternately, a pump control valve (Figure 5.3). The critical cycle period is determined, and the required closing time is directly proportional to the pipe length.

After pump start-up, the closed discharge valve is opened at a controlled slow rate, which must be field adjustable.

Because the pump is operating at near-zero flow and shutoff head, it is important while the valve slowly opens or closes that sufficient flow be available as soon as possible to relieve the high radial thrust condition inherent in the design of a volute casing centrifugal pump when operating at flows. The pump manufacturer should be consulted on minimum flow.

Under normal shutdown procedures, the valve is closed at a predetermined and field-adjustable rate before stopping the pump. Slowly opening and closing check valves or pump control valves are an effective way of controlling pressure surges during normal pump starts and stops. However, on a power-failure-caused pump trip with the sudden interruption of flow, upsurges and downsurges can occur. Valve actuators (pneumatic or hydraulic) should be provided with a reserve fluid supply to complete an emergency shutdown cycle.

AIR EXHAUST AND AIR INLET VALVES. In certain stormwater and wastewater pump discharge line configurations, the designer may elect to install valves that will admit or discharge air to the line as discharge line pressure fluctuates above and below atmospheric pressure such as during a water hammer event (Figure 5.3b).

The original filling of the line should be done in a controlled flow manner to prevent large pressure upsurges and downsurges. After filling, entrained air will come out of solution and gather at high points in the line. Air release valves located at the high points will eliminate the air, which could cause increased head loss if trapped.

The air inlet requirement to prevent vapor cavity formation should be determined for each high point. A properly sized air inlet valve should be installed to work with the smaller air release valve. These two functions may be available in a single-valve body, depending on size requirements.

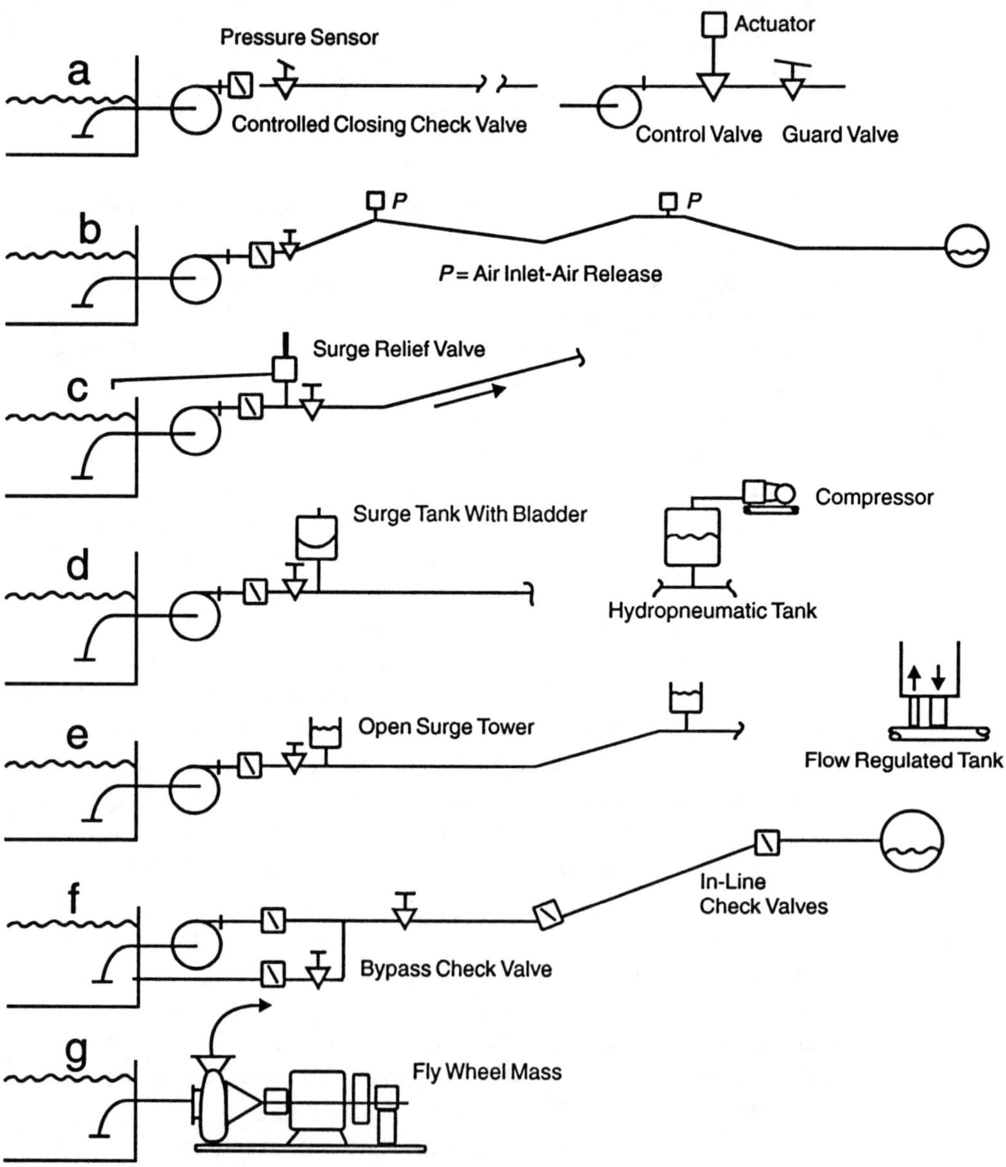

Figure 5.3 Pumping station surge reduction methods.

PRESSURE SURGE RELIEF VALVES. To relieve excess surge pressures in a pipeline, a pressure surge relief valve can be installed on the pump discharge pipe (Figure 5.3c). The valve opens on pressure above its set point and discharges flow back to the wet well. This valve is sometimes installed as a supplementary device to other water hammer control equipment.

Sizing of the surge relief valve is determined from the pipe velocity and size. A large percentage of the flow rate before flow interruption may have to be relieved to reduce pressure surge sufficiently. The valve may not respond fast enough, so a surge anticipation valve and control should be used.

Valve manufacturers have developed methods of sizing valves for specific applications. These methods normally produce a conservative design.

CLOSED OR PRESSURIZED SURGE TANKS. A closed surge tank consists of a code vessel connected to the pipeline with a compressed air source to maintain a predetermined pressure in the tank (Figure 5.3d). On upsurges the tank accepts liquid from the line and on downsurges passes liquid to the line. Smaller tanks up to 2 m^3 (70 cu ft) are available with pressurized bladders.

Surge tanks may be used in conjunction with other devices such as slow-closing and slow-opening check valves and surge relief valves. The size and location of pressurized surge tanks is best left to an expert. The design of open or enclosed surge tanks for wastewater may potentially be impacted by the solids and corrosive agents in the wastewater.

SURGE TOWERS. Surge towers are open vertical tanks with the water surface exposed to atmospheric pressure and the tank bottom connected to the pipeline (Figure 5.3e). The tank acts as an equalizer for flow variations, discharging liquid with a downsurge in the pipe or filling in case of an upsurge, though odor may be a problem. The hydraulic grade line of the discharge line must be low enough for an open tank to be used, and if needed, a surge tower at a peak in the discharge pipe profile can be constructed.

If the surge tank is located close to the pumping station, it can protect the pump on start-up from discharge line inertia that may cause the pump to otherwise operate at its zero flow or shutoff pressure, a potentially damaging condition. The height of the tower must account for the inertia head pressure, the static and friction head losses, and the upsurge head at the maximum combination.

There are several types of surge towers. By installing a check valve to allow flow only out of the tower, a downsurge can be attenuated. An altitude valve can control the filling rate to prevent overflow or, hydraulic grade line permitting, a small open return line can be installed.

BYPASS CHECK VALVES. A simple arrangement for protecting a pipeline against downsurge pressures is to install a bypass check valve in parallel with the pump (Figure 5.3f). The check or nonreturn valve would discharge in the same direction as the pump. Under normal conditions, the pump discharge pressure would be higher than the suction pressure, and the pressure difference would maintain the check valve in a closed position.

Should the pumps stop, the pressure in the discharge pipe would drop as the liquid continues to flow. If the pressure drops to below the suction pressure, water would be drawn through the bypass valve until discharge pressure and suction pressure are equal, reducing the upsurge wave pressure.

This method of surge pressure protection can only be used if the discharge pressure were to drop below the suction pressure. If the system discharge pressure is low and the discharge line velocity is high, as in a long, level discharge pipe with high flow, bypass check valves can be an inexpensive method of surge reduction. The suction reservoir level and volume must fill the pipe to avoid water column separation in the discharge line. Pressure surge relief valves may still be needed as supplementary devices.

Water could be drawn through the pump during the period the delivery head is below the suction head, which is not desirable. A controlled closing pump check or isolation valve would control this problem but could also remove the need for the bypass check.

Check valves have been placed in a longer discharge line as in-line check valves. The thinking behind this type of design is that surge pressures would be reduced, though many disagree with this viewpoint. Installing a check valve open to the direction of flow and restricting flow completely in the opposite direction could have the effect of increasing surge pressures and the potential for water hammer damage. If, as is common, a surge relief valve for power outage surge control is upstream of the check valve, the check valve restriction could limit the effectiveness of the surge relief valve. The physical system dimension and flow parameters would control the desirability and location of any in-line check valves.

PUMP AND DRIVER INERTIA. The combined rotational inertia of a centrifugal pump impeller shaft and motor rotating element would allow the system to continue rotating after power failure, delivering liquid to the discharge line, and alleviating the downsurge on the water column. However, this inertia effect seldom effectively reduces a surge event. In some systems, it may be cost effective to use a separately mounted or motor-mounted flywheel to achieve the desired inertia effect (Figure 5.3g).

SUCTION PIPING ARRANGEMENTS

Piping layouts are of prime importance to the successful operation of pumping stations. Most pump problems turn out to be related to suction piping arrangements or sump, wet well, clear well, or forebay design.

All aspects of the station hydraulic and piping designs should be reviewed carefully before design completion. All sump design features should be analyzed for potential problems, including consideration of the need for hydraulic modeling.

COMMON SUCTION PIPING PROBLEMS. The most common problems with suction piping are size, length, fittings, slope, suction side submergence, air venting, and air entrainment. Desirable suction line velocities should be 1.2 to 1.8 m/s (4 to 6 ft/sec) and less than 3 m/s (10 ft/sec). Solids deposition in lines can occur at velocities less than 0.6 m/s (2 ft/sec).

Separate suction lines to each pump should be straight and reducing fittings should be eccentric and installed with the flat side up to prevent air entrapment. Valves should not be directly connected to the pump suction or they might generate turbulence and interfere with pump operation. Piping should be properly supported and brought to the pump to prevent the transfer of piping loads to the pump casing.

Friction losses at the entrance and through the suction side piping reduce the available pressure at the pump suction. In open wet well pumping station designs, the pumps depend on atmospheric pressure and the free liquid level to bring the liquid to the pumps. Each pump requires a minimum suction pressure to pump at a given flow rate, which increases as a power function (square) of the flow rate. This minimum pressure is called net positive suction head required ($NPSH_R$) and is defined as the pressure above the liquid vapor pressure available at the pump suction. The net positive suction head available ($NPSH_A$) must exceed the $NPSH_R$. An acceptable rule of thumb for centrifugal pumps is that the pump manufacturer's guaranteed $NPSH_R$ must be exceeded by the available $NPSH_A$ by 0.6 m (2 ft).

Air entrainment in suction side piping typically is related to wet well design and the arrangement of inlets, is more common than normally thought, and causes considerable problems. If air is added to the wet well for odor control, pump air binding may be a problem.

The practice of sloping wet well walls toward the suction inlet should be studied carefully, because it is a common source of air entrainment. Flushing the wet well to remove sediment by hosing is an alternative. If the wet well level elevation is below the packing of a dry-pit pump without a water seal, the force of atmospheric pressure may force air into the pump, causing the pump to air bind and lose prime. To avoid problems, the low wet well liquid level should be above the centerline of the pump impeller, called a flooded suction.

STATION PIPING

Smaller pumping stations have two or three pumps, typically of the same size, and larger stations may have more. The discharge piping of the pump to the station discharge is termed station piping.

There are numerous methods of affecting discharge from a pumping station and some general rules that should be followed. As mentioned in Chapter 3, the pump discharge check valve should be mounted in a horizontal plane, not vertically where solids may accumulate on the valve disk. Flow to a common discharge header from pump piping should be to the side of the header, not the bottom, to avoid entrained solids from settling to the discharge pipe of an inactive pump. Entering the discharge header at an angle, as from a lateral fitting, can reduce minor fittings losses.

DISCHARGE LINE PIPING

The discharge line from the pumping station should be considered an integral part of the system. Suction, station, and discharge piping determine the system head curve, critical to proper pump selection.

The discharge line diameter should provide a minimum line velocity of 0.6 m/s (2 ft/sec). Line velocity in excess of 2.5 m/s (8 ft/sec) should be avoided except on very short lines. All piping must be supported for weight and for lateral and axial thrust forces generated by internal pressure and changed direction of flow.

With constant-speed (intermittent pumping) design, the discharge pipe liquid must be accelerated from rest at each pump start. The designer should check for surge and acceleration head requirements for each design. Designs with a high proportion of static head to total head and those with longer force mains should be reviewed for the acceleration head effect. From Newton's second law, a relationship can be developed for the elapsed time to accelerate the liquid in the discharge pipe from zero velocity to design velocity:

$$t = g^{-1} L \cdot V / (H\text{so} - H\text{s}) \qquad (9)$$

Where

L	=	discharge line length, m (ft);
V	=	velocity, m/s (ft/sec);
Hso	=	centrifugal pump zero flow head, m (ft);
Hs	=	system static head, m (ft);
g	=	gravitational constant, m/s^2 (ft/sec^2); and
t	=	pause time to accelerate, seconds.

In operation, the centrifugal pump starts and develops the shutoff or zero flow pressure until acceleration of the liquid in the line permits the pressure to drop to the normal intersection of the system head curve with the pump performance curve. The pause point in the start-up sequence is at a flow rate less than the minimum noted by the pump manufacturer. Some manufacturers do not show minimum and maximum operating flow rates on their performance curves.

All volute casing centrifugal pumps develop high radial thrust forces as they operate close to their zero flow (shutoff head) condition (Hydraulic Institute Standards, 1983). These high radial forces impose reverse bending loads on the shaft of overhung impeller pumps, which may produce stresses in excess of the endurance limit of the shaft material, resulting in early fatigue failure of the shaft. The loads are also transmitted to the pump bearings, causing them to fail. The designer should calculate the parameters for his application. Generally, if the pause time exceeds 10 seconds, the design should be modified and the pump manufacturer should be consulted for ways to avoid this problem.

Pumps are started in some applications against a closed valve, which should be programmed to start opening when the pump reaches full speed. Variable-speed drive systems should start or operate pumps within the recommended operating range for the speed choices.

All discharge lines should be given an adequate analysis to ensure that surge pressures will not adversely affect system operation. This item is covered earlier in this chapter.

For underwater crossings and similar extended low points or inverted siphons, it is advisable to design for periodic cleaning or equip the line for blowoff to a collecting manhole.

PIPING MATERIAL AND FITTINGS

The most common pipe materials used in wastewater and stormwater pumping stations are ductile iron and steel. Plastics are used for discharge lines 200 mm (8 in.) and smaller in diameter. Concrete pressure pipe also is common on discharge lines outside the pumping station in larger sizes. Composite piping materials are also used. The selection of a pipe material is usually based on pipe strength, corrosion resistance, local regulations, and cost.

Suction side piping and fittings in dry well–wet well station piping typically are of ductile iron. Discharge station piping and fittings are either of steel or ductile iron construction. The ability to fabricate complex steel shapes for station discharge piping has increased the use of steel. Steel pipe must be protected against corrosion both externally and internally.

Cast iron pipe fittings are available but more brittle than those of ductile iron. Piping and fittings of iron can be internally coated with polyethylene, asphaltic, or cement mortar linings for corrosion protection and to maintain a low friction loss coefficient.

PIPE JOINTS AND CONNECTIONS. Steel or iron piping sections typically are joined by gasketed flanged joints 100 mm (4 in.) in diameter and larger. For the pressure application in wastewater and stormwater stations, class 55-kg (125-lb) drilling pattern flat face flanges commonly are used. Class 110-kg (250-lb) flanges are also available. Screwed coupling and fittings are used in smaller sizes to 100 mm (4 in.) and lines sometimes 150 mm (6 in.) in diameter. Pipe flanges may be integrally cast or threaded to the pipe by the manufacturer.

There are several proprietary clamp-on joint fittings, which permit a level of deflection in the pipe and simplify installation and future repair or replacement. Grooved pipe and pipe fittings are used to provide leak-tight connections while allowing for some angular pipe deflection.

Sleeve or sleeve-type couplings are used to join pipe and allow axial expansion and misalignment. It is important that the joints be harnessed with restraining bolts and provided with restraints at bends.

Pipe for underground force mains should be supplied with bell and spigot joints when outside the pumping station. To avoid or reduce the differential settling effect between the pumping station and the force main, a flexible pipe joint may be installed outside the station between station piping and the force main. Mechanical joint fittings, which use bolted retainer glands, are also used in underground piping when the joints must be tied.

PIPE SUPPORTS FOR STATIC LOADS. Piping systems must be structurally sound. The static support system must consider the structural strength and weight of the pipe and fittings and the jointing arrangement. Thermal expansion of the pipe must be recognized.

Static supports of concrete or metal saddles and hangers are used to support the weight of station piping. Base-mounted elbows frequently are used along with lateral supports on vertical pipes. The spacing of supports and their application have been calculated and presented by the manufacturers (Hanger Standards, 1987). The design of concrete supports should provide for axial pipe movement without progressive wear of the pipe exterior wall at the support.

Vibration from surge pressures or other hydraulic transients imposes unusual loads on pipe supports, making it prudent to design conservative support systems and provide for both weight and axial-lateral loads. Support systems must be designed for possible overpressurizing in required pipe testing. Computer programs, such as Surge—A Computer Analysis of Transient Flow, are available for some or all of this work.

Expansion joints, sometimes used at pump flanges, typically are constructed of two metal flanges with a single arch nonmetallic tube connecting the flanges. The use of expansion joints with proper anchoring devices will reduce, but not eliminate, vibration to or from the pumps and piping system. Specifications should call for proper expansion joint installation (Technical Handbook, 1980), though their use should not be an excuse to use the fitting to accommodate misaligned piping.

An expansion joint should be installed with an anchor point between the expansion joint and the pump. The designer should realize that exposed joints must be harnessed or blocked to resist large thrust forces developed with the expansion joint. Figure 5.4 illustrates one method of avoiding the transference of these forces to the pump (Technical Handbook, 1980).

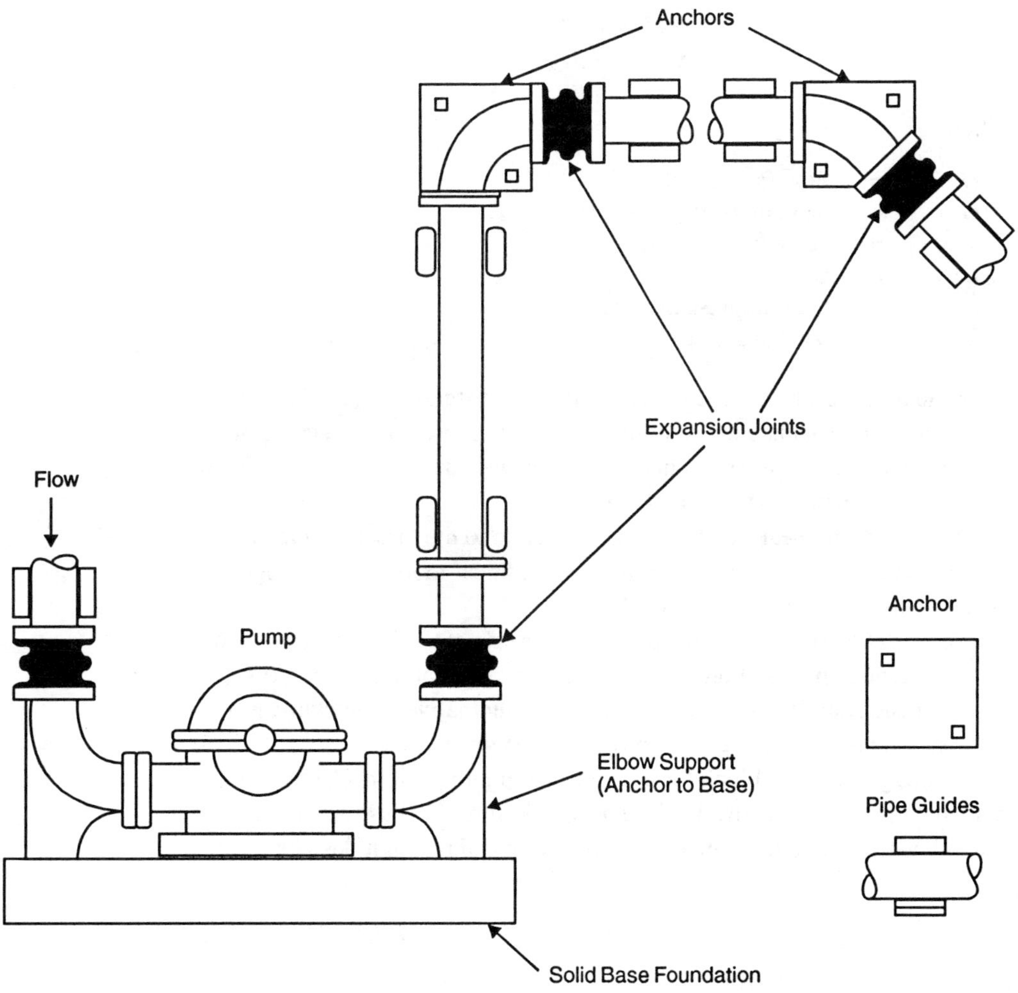

Figure 5.4 Typical piping layout using expansion joints when equipment and piping are properly anchored.

PIPE SUPPORTS FOR DYNAMIC LOADS. When a pipe filled with liquid encounters a change in direction, as at a bend, unbalanced thrust forces (pressure and momentum forces) must be balanced with anchored supports called thrust blocks. Thrust blocks for underground pipes typically are concrete structures designed for the thrust loads and soil conditions. In-station piping must also be analyzed. The basic relationships for treating these loads are shown below.

A summation of the forces relating to internal pressures (*R*) and the forces relating to changes in linear momentum (*M*) are

$$\begin{aligned} R_x &= P_1A_1 - P_2A_2 \cos\theta \\ R_y &= P_2A_2 \sin\theta \\ M_x &= GQ\,[V_2 \cos\theta - V_1] \\ M_y &= GQ\,[V_2 \sin\theta] \\ R^2 &= R_x^2 + R_y^2 \end{aligned} \quad (10)$$

Where

Q	=	flow rate (constant);
G	=	mass density (constant);
V	=	velocity;
A	=	area;
P	=	internal pressure; and
θ	=	bend angle degrees.

In the most general case, areas A_1 and A_2 are different. Figure 5.5 shows the formulas to calculate the reaction load for the most common cases but does not consider momentum changes. The largest area of a bend should be used to yield a conservative thrust value.

The change in linear momentum contribution to the thrust load reaction is minimal in most cases. The contribution from linear momentum changes increases with velocity change of the fitting.

For in-station piping, thrust restraints are made of steel. Concrete is typical for underground piping. Figure 5.6 shows some general in-station thrust blocking and restraining arrangements. As an alternative to blocking, restrained joint piping can be used to resist thrust forces.

For design of thrust block for underground piping, soil mechanics and pipe material determine much of the design. Figure 5.7 illustrates underground, vertical, and horizontal thrust blocking and restraint designs.

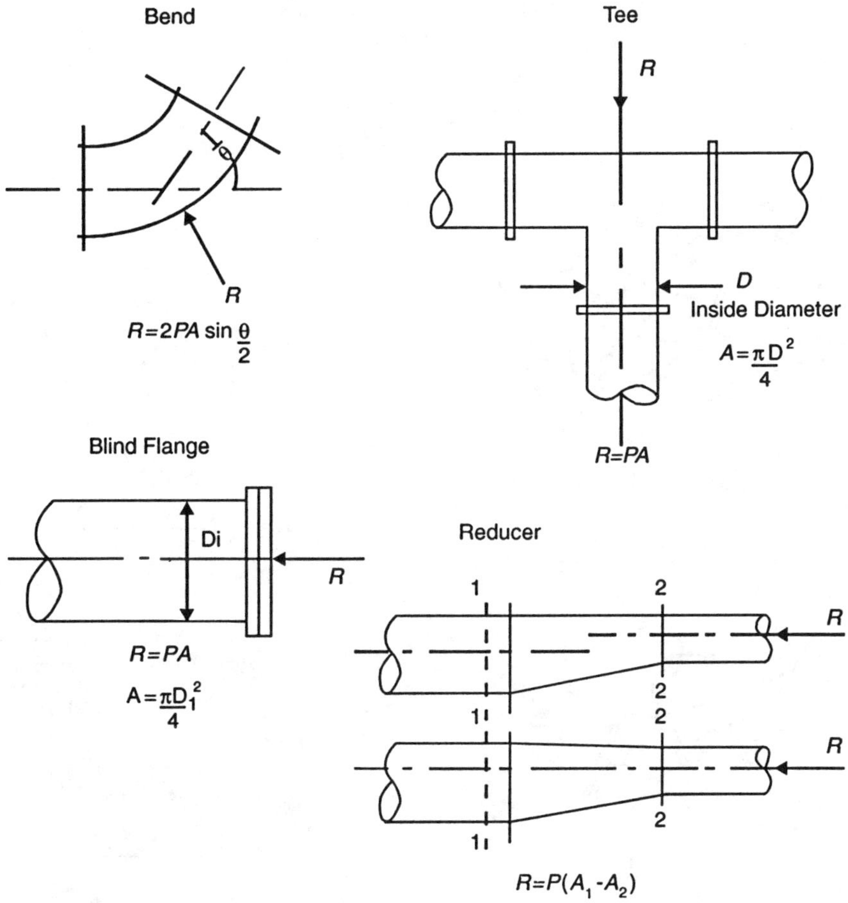

Figure 5.5 Pipe thrust force formulas.

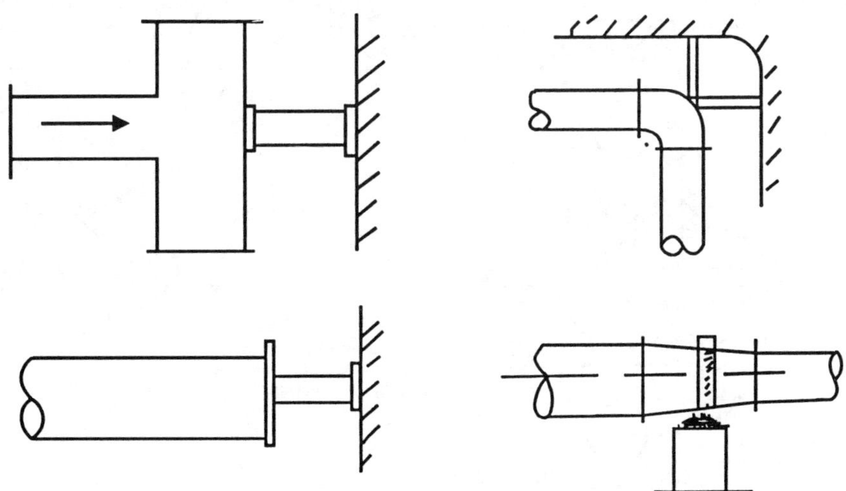

Figure 5.6 Station piping thrust restraints.

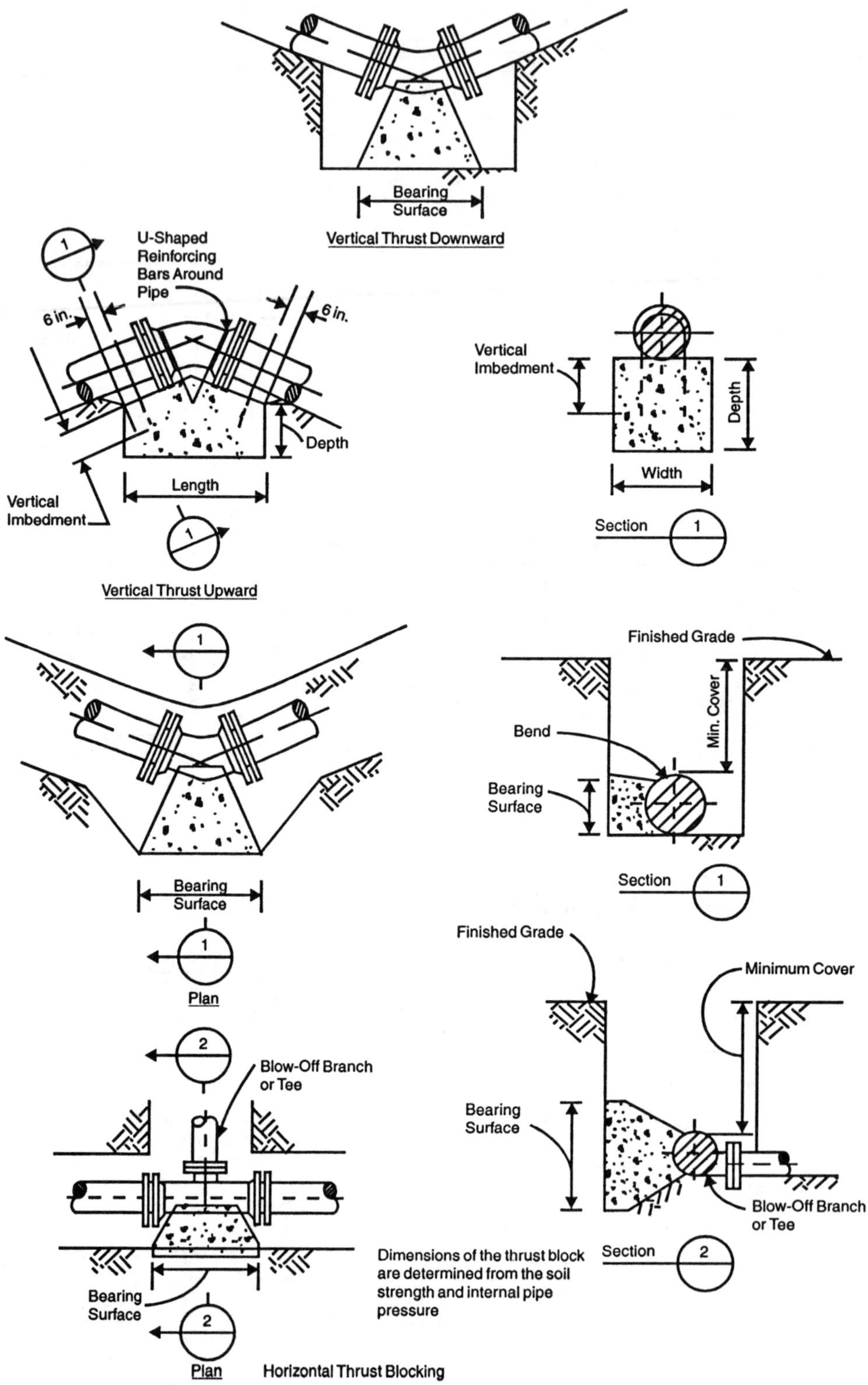

Figure 5.7 Underground piping thrust restraints (in. × 25.4 = mm).

VALVES

Pumping stations use valves to isolate a pump, prevent backflow through the pump, isolate the valve that prevents backflow, divert or bypass flow, isolate segments of lines for servicing, and surge control. Figure 5.8 shows a general schematic for a typical pumping station valving arrangement.

VALVE TYPES. Valves that readily clog, such as butterfly valves (Figure 5.9), should not be used with stormwater and wastewater, which contain indiscriminately sized solids. Generally, where clogging can occur, plug or ball valves (Figures 5.10 and 5.11) typically are used.

Pinch valves (Figure 5.12) sometimes are effective as flow control devices. Valves are used to isolate lines and permit repair of pipe or fittings or other valves. Sluice gates (Figure 5.13) are used to seal off wet well sections.

The operation of a cone valve (Figure 5.14) is unique in that it involves lifting the cone, which then rotates 90 degrees to open the valve. Cone valves are used for isolation and surge relief, especially in larger sizes. A guard or isolation valve is placed downstream to permit maintenance of the cone valve. Regular exercise of the valve is important. Cone valves with hydraulic or electric actuators are used for control valves.

To prevent the return flow of a liquid stream, check valves are used. The two principal types are swing check valves (Figure 5.15) and ball check valves. Pinch check valves (Figure 5.16) are used to partially isolate larger gravity stormwater outlets and to prevent the entrance of small animals.

Check valves should be installed in a horizontal position. Because the closing rate of the check valve is a function of the hydraulic system design,

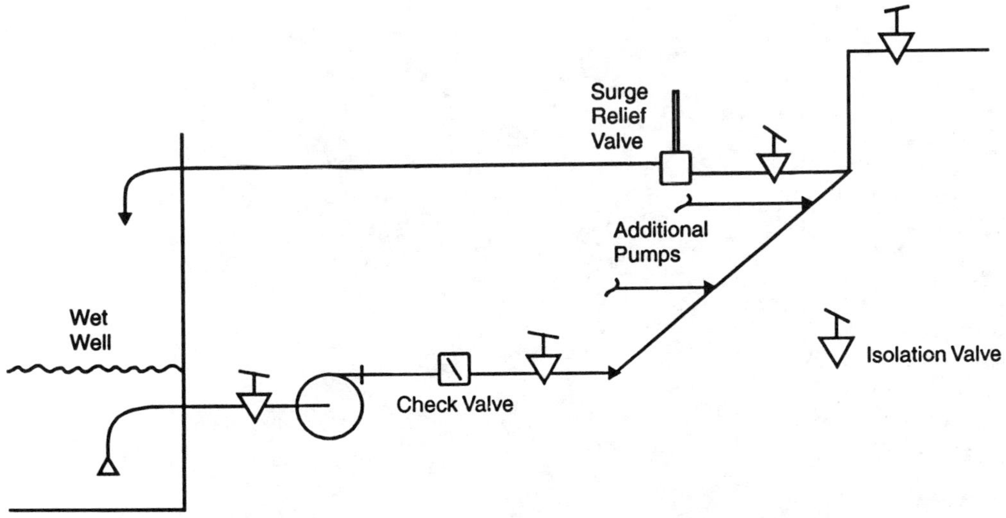

Figure 5.8 Pumping station valve locations.

various devices, springs, and weights are used to control check valve closing. Controlled closing check valves (refer to Figure 5.15) are also used on occasion to prevent slamming.

Relief valves are used to relieve excess pressure in pipes and are typically of an angle globe valve design with spring adjustment for regulating pressure (Figure 5.17).

Air release and combination air and vacuum valves can be placed at high points to release accumulated air or admit air under downsurge operating conditions. The valves used in wastewater (Figure 5.18) are of different design than those used for clearwater service.

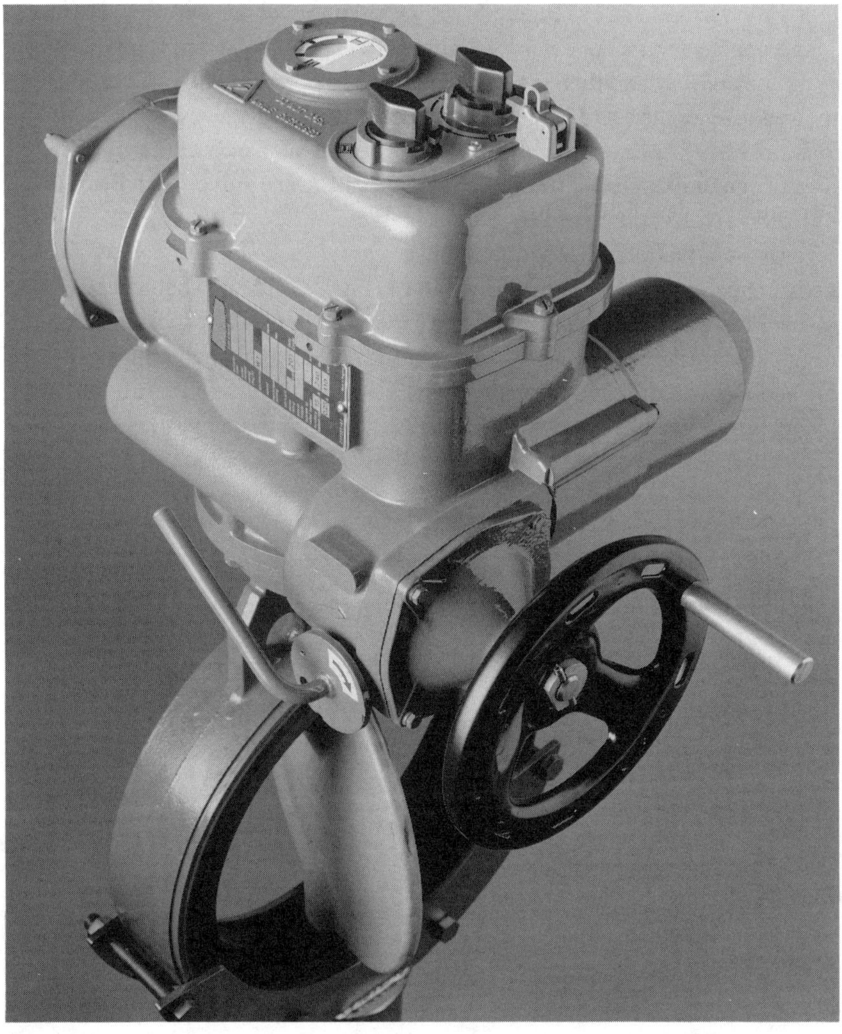

Figure 5.9 Butterfly valve. Courtesy of Rotork Controls Inc.

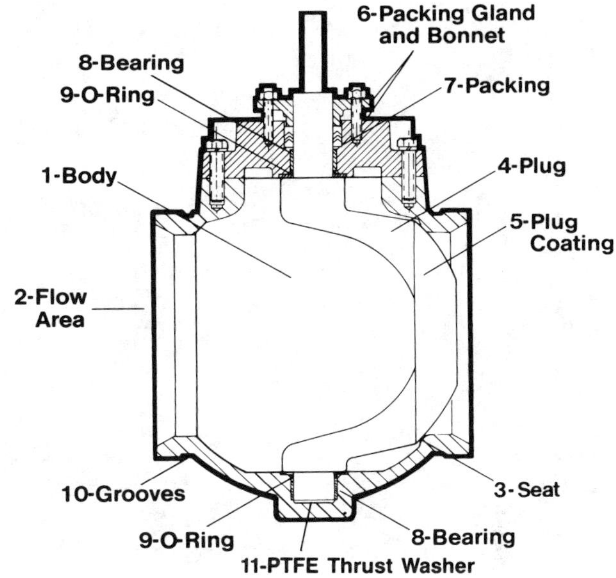

Figure 5.10 **(a) Eight-in. plug valve and (b) interior schematic. Courtesy of Victaulic Company of America.**

Figure 5.11 **Two-in. ball valve. Courtesy of Victaulic Company of America.**

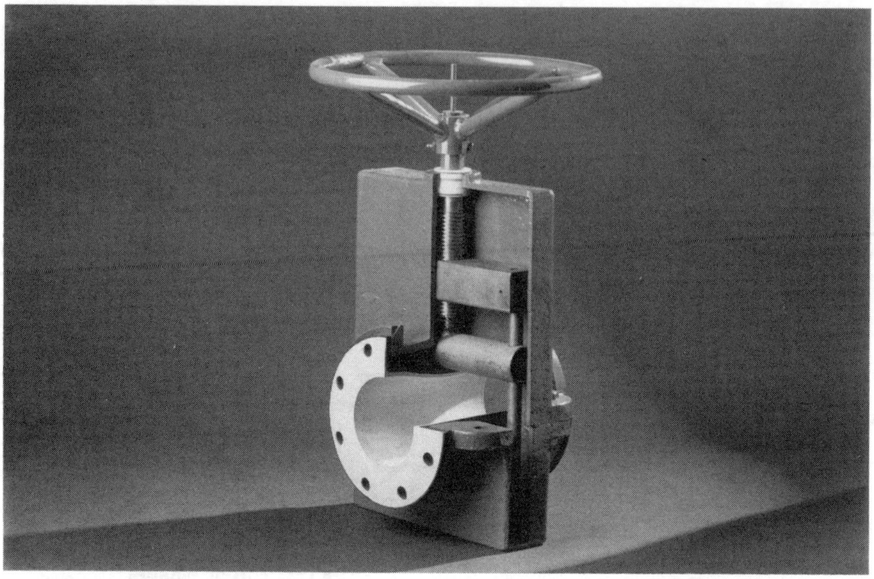

Figure 5.12 Manual control pinch valve. Courtesy of Red Valve Co. Inc.

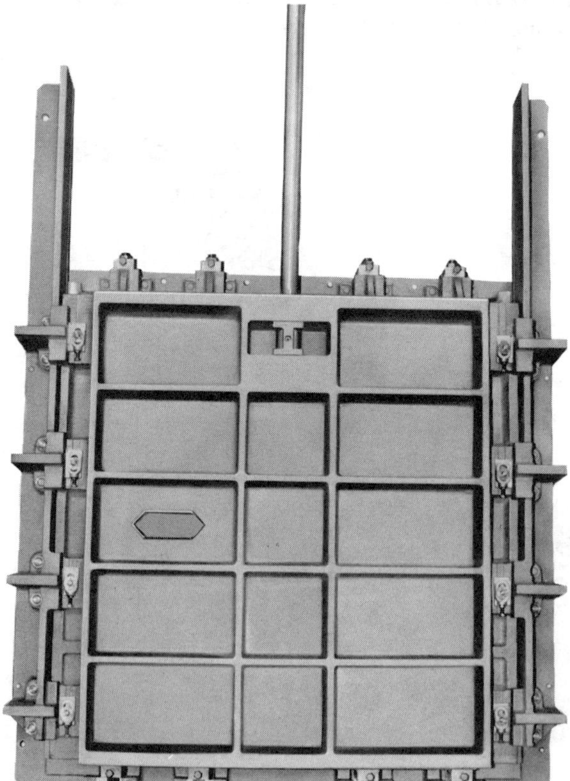

Figure 5.13 Sluice gate. Courtesy of Rodney Hunt Co.

Trapped air at pipeline high points can increase power requirements and the potential for pressure surges. Manual air release valves are a possible solution to this problem.

Trapped gases can be high in hydrogen sulfide and cause considerable problems with odors. Also, the orifice size in the valve may cause high velocities, which can produce elevated sound levels in the vicinity. Smaller accessory valves may be of globe or gate design.

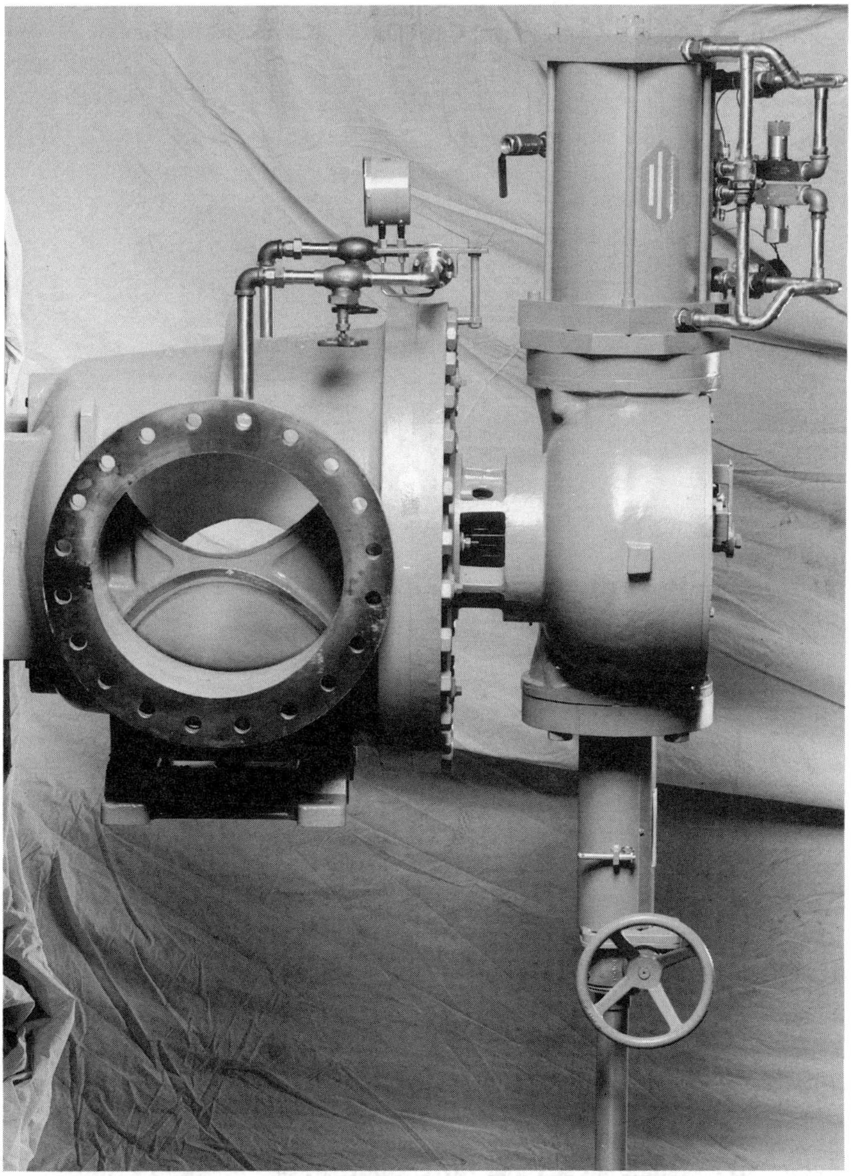

Figure 5.14 Cone valve (partially open).

Figure 5.15 Swing check valve. Courtesy of Valve and Primer Corporation.

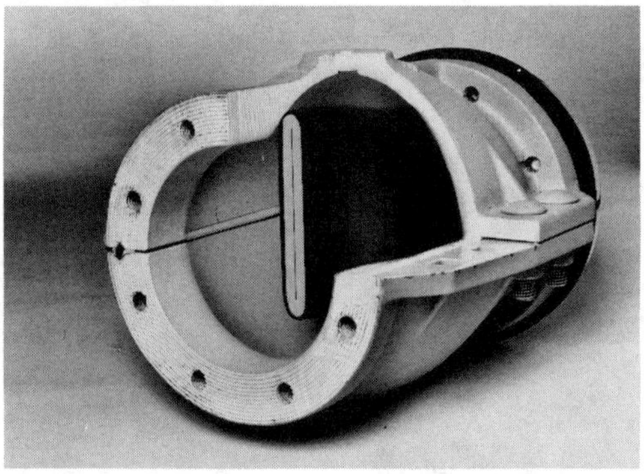

Figure 5.16 Pinch check valve. Courtesy of Red Valve Co. Inc.

Figure 5.17 Angle surge relief valve. Courtesy of Valve and Primer Corporation.

Figure 5.18 Wastewater combination air valve. Courtesy of Valve and Primer Corporation.

VALVE ACTUATION. Smaller valves can be opened and closed manually by adjustment of an operator (refer to Figures 5.9, 5.10, and 5.12 as examples). Line pressure and valve size may require the use of mechanical actuation devices (gears and chain drives) to facilitate opening and closing.

Powered actuators are used for larger valves to provide remote operation or positively open or close valves. Actuators are powered by electric, pneumatic (compressed air), or hydraulic pressure (Figures 5.19, 5.20, and 5.21,

Figure 5.19 Electric valve actuator. Courtesy of Rotork Controls Inc.

respectively). Actuators can simply open or close a valve or be used to position a valve. The rate of opening and closing can be made adjustable.

Pumping systems frequently use power-actuated valves as an isolating device. Where ordinary swing-type check valves are not suitable, controlled closing and opening devices may be installed on the swing check. A plug or ball valve can also be used.

Figure 5.20 Pneumatic (air) valve actuator. Courtesy of Rotork Controls Inc.

Figure 5.21 Hydraulic valve actuator. Courtesy of Rotork Controls Inc.

VALVE APPLICATIONS. In wastewater and stormwater pumping stations, valves are used to interrupt or control flow, vent gases or relieve a vacuum condition, or relieve excess pressure from a system.

An important application is pump check service. During a power failure, the actuator (pneumatic or hydraulic) can operate the check valve in a predetermined emergency cycle if supplied with emergency air or liquid under pressure.

Valves should not be mounted directly on the suction side of a pump as the valve port can create turbulence in the pump suction, leading to serious problems.

Surge relief valves should be sized with minimum pressure drop in the discharge line. Some installations use two surge relief valves, frequently of different size, operating in parallel with slightly different pressure-relief settings. The port area of the two valves must equal a port area of the larger single valve required.

FLOW METERS

In addition to establishing flow rate for billing and remote flow control, flow is measured at pumping stations to test pumping systems precisely. Even under the best planned conditions, field tests are difficult to perform accurately. A properly installed and recently calibrated flow meter can add accuracy to the field test.

REFERENCES

Cameron Hydraulic Data (1992). 17th Ed., Ingersoll Rand Co., Woodcliff, N.J.

Flow of Fluids (1974). Technical Paper No. 410, Crane Co., New York, N.Y.

Hanger Standards (1987). Grinnell Corp., Providence, R.I.

Hydraulic Institute Engineering Data Book (1979). Hydraul. Inst., Cleveland, Ohio.

Hydraulic Institute Standards (1983). 14th Ed. Hydraulic Inst., Cleveland, Ohio.

Krebs, J.R. (1990). KCS Pump—Centrifugal Pump Selection by Computer. Krebs Consult. Serv., Dayton, Ohio.

KY Pipes, Univ. of Ky. Dept. of Civ. Eng., Lexington. Computer Program.

Micro Hardy Cross, CEComp, Granada Hills, Calif. (1986) Computer Program.

Parmakian, J. (1963) *Water Hammer Analysis*. Dover Press, New York, N.Y.

Pump-Flo: A Pump Selection Program, Engineered Software, Olympia, Wash.

Recommended Standards for Waste Water Facilities (1990). Health Educ. Serv., Albany, N.Y.

Submersible Sewage Pumping (1986). Lewis Publishers, Chelsea, Mich.

Surge—A Computer Analysis of Transient Flow, Univ. of Ky. Dept. of Civ. Eng., Lexington.

Technical Handbook (1980). 5th Ed., Fluid Sealing Assoc. Rubber Expansion Joint Div., Philadelphia, Pa.

Water Pollution Control Federation (1979). *Odor Control for Wastewater Facilities*. Manual of Practice No. 22, Washington, D.C.

Wood, D.J., *et al.* (1991) Waternet-Flow Distribution Network Analysis. Univ. of Ky., Dept. of Civ. Eng., Lexington.

SUGGESTED READINGS

Chaudrey, M.H. (1987) *Applied Hydraulic Transients.* Van Nostrand Rheinhold, New York, N.Y.

Design Recommendations Pumping Stations with Large Submersible Centrifugal Pumps (1993) Flygt Corp., Norwalk, Conn.

Karassik, I.J. *et al.* (1976) *Pump Handbook.* McGraw–Hill, Inc., New York, N.Y.

Krebs, J.R. (1993) *Pump Selection and System Design with the Microcomputer.* Lewis Publishers, Chelsea, Mich.

Metcalf and Eddy, Inc. (1981) *Wastewater Engineering: Collection and Pumping of Wastewater.* G. Tchobanoglous (Ed.), McGraw–Hill, Inc., New York, N.Y.

Prosser, M.J. (1977) The Hydraulic Design of Pump Sumps and Intakes. Br. Hydrodynamic Res. Assoc., Cranfield Bedford, U.K.

Streeter, V.L., and Wylie, E.B. (1967) *Fluid Transients.* FEB Press, Ann Arbor, Mich.

Thorley, A.R.D. (1992) *Fluid Transients in Pipeline Systems.* D&L George, Ltd., Barnet Herts, U.K.

U.S. EPA (1979) Sludge Treatment and Disposal. EPA-625/1-79-011, Cincinnati, Ohio.

Chapter 6
Electrical Design

Electrical design of wastewater and stormwater pumping stations must account for all electrical loads and provide cables and service equipment that meet local standards and codes. It also must provide cables for interfacing the components of the instrumentation and control system.

An electrical system can range from a simple fractional kilowatt motor plugged into a convenience outlet to a complex group of pumps with megawatts of connected load. Local jurisdictions have final authority for the enforcement of standards and codes used in electrical system design.

RELIABILITY AND REDUNDANCY

Reliability and redundancy requirements of electrical system design are covered in the codes and standards. The National Electrical Code (NEC) covers requirements for the electrical system conductor sizing, grounding, motor circuit protection, installation requirements for hazardous areas and installation requirements for standby generation (National Electrical Code, 1993). The National Electrical Safety Code covers requirements for substation and overhead line designs (National Electrical Safety Code, 1990). The National Fire Protection Association (NFPA) 820 defines hazardous areas for pumping stations (Recommended Practice, 1990). The U.S. Environmental Protection Agency's "Technical Design Criteria for Mechanical, Electrical, and Fluid System and Component Reliability" (1974) provides design guidelines for electrical system redundancy.

The Institute of Electrical and Electronic Engineers Incorporated (IEEE) Standard 493 covers the recommended practice for reliable electrical system design, providing information on electrical equipment failure rates and actual hours of down time for a failure (Gold Book, 1990).

The local jurisdiction has final authority on standards used in electrical design for any project and must be consulted about their requirements for the electrical system. Also, the means of service to the facility by the local utility company must be incorporated into the design.

The utility's reliability is reflected in its outage history for the service line to the facility: what happens if power to the facility fails, and what facilities does the owner or utility have for handling such an occurrence? An outage will not be a problem if storage capacity is built into the system or if there is available standby pumping via portable pumps or standby generators.

For other facilities, provisions are needed to ensure pumping is not lost. The level of reliability and redundancy needed will be dictated by the political realities of the facility's location. The installation of a standby pump sometimes will be sufficient for some facilities. For others, standby generation or a second utility source will be needed. In critical locations, the facility may require redundant backup sources of power, redundant means of electric- and engine-driven pumping, and means to connect portable pumps or standby generation.

How an electrical system is designed and the equipment installed also will affect the reliability of the system. The NEC in Article 110 covers spacing requirements for electrical equipment. The NEC in Article 210, part C covers requirements for receptacle locations in dwelling units, but not for receptacles in industrial areas such as pumping stations (National Electrical Code, 1993). Receptacle locations for hand tools or welding outlets need to be addressed in the electrical design to ensure proper facility maintenance.

Manufacturer's standards for electrical equipment being installed contain limitations on temperature, humidity, altitude, voltage, interrupting ratings, and clearances.

With the installation of more electronic controls in pump stations, grounding of the station will affect system reliability. Station grounding is covered by the NEC in Article 250, and IEEE standard 142 (Green Book, 1982) is a guide for grounding design. Information on another aspect of system grounding, surge protection, can be found in IEEE Standard C62 (Guides and Standards, 1990).

VOLTAGE CONSIDERATIONS

The service voltage is chosen after reviewing the facility's loads for connected kilowatts, running kilowatts, and largest individual kilowatts. Contact the utility company about available voltages and requirements for service.

The utility company determines what voltage can be supplied to the site based on the requirements of their system. Nominal system voltages are listed in Table 6.1. The station designer determines the voltage by the power rating of the motors needed to drive the pumping equipment and selects the best standard voltage rated equipment. Motor standard voltage ratings are in Table 6.2.

Table 6.1 Nominal system voltages.

	Single phase	Three phase	
Voltage class	**Three wire**	**Three wire**	**Four wire**
Low voltage	120/240 V		208 Y/120 V
		240 V	240 /120 V
		480 V	480 Y/277 V
			600 V
Medium voltage		2 400 V	
		4 160 V	
		4 800 V	
		6 900 V	
			12 470 Y/7 200 V
			13 200 Y/7 620 V
		13 800 V	
			13 800 Y/7 970 V

Table 6.2 Alternating current motor standard utilization voltages.

Single phase	Three phase
115 V	200 V
230 V	230 V
	460 V
	575 V
	2 300 V
	4 000 V
	4 600 V
	6 600 V
	13 200 V

The power rating ranges for standard motors shown in Table 6.3 represent a manufacturer's typical range for motor construction.

Table 6.3 Manufacturer's standard power ranges per voltage class.

Single phase	
Voltage rating	**Power**
115 V	Fractional to 10 hp
230 V	Fractional to 10 hp
Three phase	
Voltage rating	**Power**
200/230/460 V	Fractional to 600 hp
2 300 V	200 to 4 000 hp
4 000 or 4 600 V	400 to 7 000 hp
6 600 V	1 000 to 12 000 hp
13 200 V	3 500 to 25 000 hp

1 hp × 0.746 = 1 kW.

Voltage must be selected with allowances for voltage drop in the circuit wiring. The distance from point of service to point of use must be considered because of the voltage drop caused by the current draw of the motor and the resistance of the wiring to that current flow. A motor can be operated without damage in a voltage range of plus or minus 10% of the motor's standard voltage rating.

Motor inrush currents caused by starting a motor must be evaluated when selecting the proper voltage. Across the line starting of a motor causes locked rotor currents six times the motor's full load rated current. These currents are caused by the electromagnetic force necessary to cause the motor shaft to turn and to bring the shaft up to speed. Motor starting currents can be a major concern of the electrical utility, if they cause flickering on its system as a result of voltage dip at the starting of the motor.

Inrush currents also affect sizing of standby generation equipment, because the generator must have the capacity to handle the motor's starting currents. These can be reduced by using reduced voltage starters, wye-delta part-winding starters, solid state, soft-start starters or adjustable-frequency drives.

The voltage class of equipment being considered needs to be evaluated, as do the training and experience of the maintenance staff. Installing the wiring for medium-voltage equipment requires special skills, and the NEC in article 110 discusses space requirements for electrical equipment (National Electrical Code, 1993). As the equipment voltage class goes up, so does the need for clear space in front of and behind the equipment.

SUBSTATIONS

Voltage selection may require the installation of a substation to provide the facility on-site transformation of the utility voltage to the equipment's voltage rating. The substation equipment can be owned and furnished by the utility or as part of the facility's equipment. A review of the rate history of the utility for metering on the primary versus the secondary side of the transformer will determine if the cost of the transformation equipment and the assumption of risk should the equipment fail are justified.

The following IEEE standards provide guidelines for the design of a substation and its equipment: (1) Standard 80, "Guide for Safety in AC substation: Grounding" (1986); (2) Standard 1109, "Guide for the Interconnection of User-Owned Substations to Electrical Utilities" (1990); and (3) Standard C37, "Circuit Breakers, Switchgears, Relays, Substations, and Fuses."

Needs for reliability and redundancy may necessitate a second electrical source at a facility, either a second transformer in a second-source scheme or on-site generation. The risk of electrical system supply failure is lessened with a two-source scheme, be it two transformation points or one source transformation point and generator. A tie breaker between the two sources is required to add further reliability to the system.

The substation design will need to be coordinated with the utility company, which will need to know the following:

- Where the metering is to be installed;
- What loads are going to be served;
- How many sources of power are required;
  - What loads each will serve; and
  - What standby capacity each source will be required to have.

The utility may require an easement for the installation of their lines and equipment on the property, and perhaps a one-time charge at the start-up of the facility. The utility company typically will recover other costs of providing service, such as transformer losses, in its rate schedule.

TRANSFORMERS

Guidelines for the manufacture, installation, and testing of distribution, power, and transformers are provided in IEEE Standard C57 (Distribution, 1990).

As stated in IEEE standard 493 (Gold Book, 1990), transformers have a low incidence of failure, but outages are among the longest for a piece of electrical equipment. Utility companies generally have spare transformers in their service yard for most standard pole units. A large-station-class unit can take months to replace if it should fail.

Transformers can have an overload capacity above the nameplate rating if forced air cooling is added to the transformer or if the transformer is specified with a higher temperature rise over standard. Transformers come in standard kilovolt ampere (kVA) ratings. See Table 6.4 for a single-phase unit and Table 6.5 for a three-phase unit (Red Book, 1986).

Table 6.4 Standard single-phase transformer kVA ratings.

3	75	1 250	10 000
5	100	1 667	12 500
10	167	2 500	16 667
15	250	3 333	20 000
25	333	5 000	25 000
37.5	500	6 667	33 333
50	833	8 333	

Table 6.5 Standard three-phase transformer kVA ratings.

15	300	3 750	25 000
30	500	5 000	30 000
45	750	7 500	37 500
75	1 000	10 000	50 000
112.5	1 500	12 000	60 000
150	2 000	15 000	75 000
225	2 500	20 000	100 000

Transformers can be divided into two basic groups by their insulating material, dry type, or liquid immersed. Both types are designed to handle their full-load rated capacity without additional cooling at an average ambient temperature of 30°C (85°F) for a 24-hour period, with the maximum temperature not exceeding 40°C (105°F).

For a dry-type transformer, the temperature of the cooling air when the unit is operating should not drop below minus 30°C (–22°F), and the altitude of the unit should not exceed 1 000 m (3 300 ft) without derating the unit.

For a liquid-immersed, air-cooled transformer operating without additional cooling, the surface temperature of the liquid should not fall below –20°C (–4°F). Similarly, the altitude should not exceed 1 000 m (3 300 ft) without derating the unit.

Atypical transformer service conditions need to be addressed in the design and layout of the equipment. Such conditions occur when ambient temperatures rise or fall or through exposure to damaging fumes, vapors, abrasive dusts, salt spray, excessive moisture or dripping water, abnormal vibration, shock, tilting, and seismic conditions.

Installation requirements for transformers are covered in NEC Article 450, American National Standards Institute (ANSI)/IEEE C57.12.01 for dry-type transformers and ANSI/IEEE C57.12.00 for liquid-immersed transformers (Distribution, 1990).

Transformer impedance is a large factor in fault studies that controls the maximum amount of fault current let through to the secondary side. The higher the impedance of the transformer the lower the fault current will be on the secondary side of the transformer.

SWITCHGEAR

Switchgear provides reliability in the facility design. By splitting the main bus of the switchgear with a tie-interrupting device, the facility loads can be divided between halves of the equipment, allowing isolation in case of equipment failure of an upstream device. The tie enables the halves to be connected, and this arrangement as illustrated by Figure 6.1 is called having a selective system. A switchgear located on the primary of a transformer is called a primary selective system. Having the tie on the secondary side of the transformer provides a secondary selective system.

Requirements for the design, installation, and testing of switchgear assemblies are discussed in IEEE Standard Collection C37 (Circuit Breakers, 1989).

There are two types of switchgear assemblies, open and metal enclosed. In open switchgear, the main bus bar, interconnecting wiring, interrupting device, associated controls, and metering are not enclosed in a supporting structure. In metal-enclosed switchgear, these items are enclosed on all sides and the top by a metal supporting structure.

Requirements for working space around this equipment are based on the equipment's utilization voltage (National Electrical Code, 1993). Installation requirements of the manufacturer cover working space required for the removal of interrupting devices for the switchgear assembly.

According to major manufacturers of switchgear rated 600 V and below with bus rated 1 600 A and below, the equipment can be mounted against the wall. At 2 000 A, the equipment will require back access to make cable terminations. Standard continuous current rating for buses of metal-enclosed switchgear is covered in Table 6.6.

The interrupting devices that make up the main component of this equipment are named for the insulating material used in their design. In medium-voltage equipment, the interrupting of the circuit is done in a vacuum, in air,

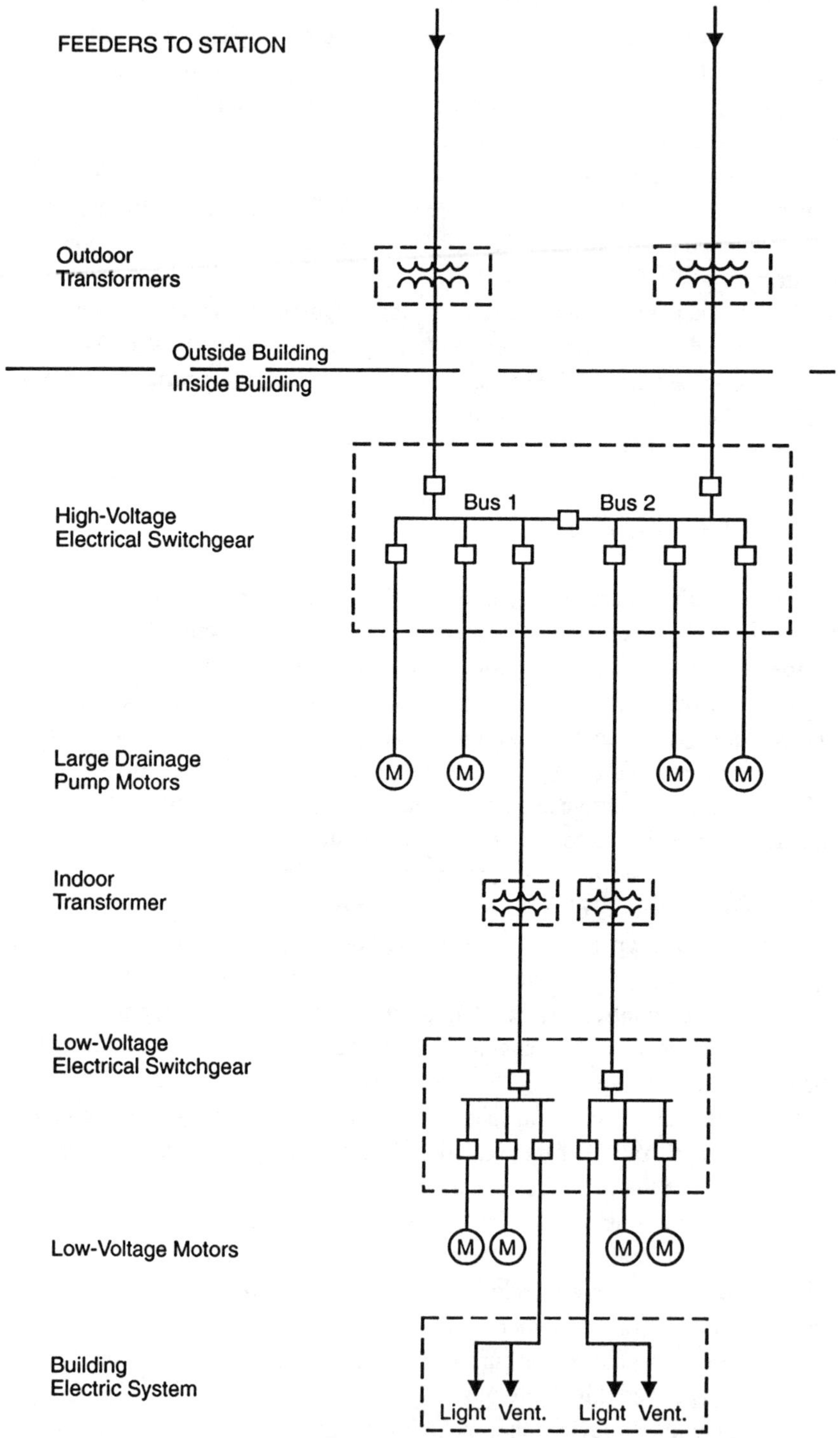

Figure 6.1 One-line diagram for a stormwater pumping station.

Table 6.6 Manufacturer's standard bus ratings by equipment voltage class.

To 600 V	From 600 to 13 800 V
600 A	—
1 200 A	1 200 A
1 600 A	—
2 000 A	2 000 A
3 000 A	3 000 A
4 000 A	—

with a gas (SF6), or with fuses. The selection of interrupting device used at medium voltage is based on the fault-interrupting capacity of the device, the speed of fault interruption required, the space required to house the device, and the cost. Coordination among interrupting devices used to be a concern, especially between systems with fuses and relaying, but electronic fuses have solved the problem of coordinating a fused switch with a relay-based protection system.

The reliability of an electrical system is affected by how well its interrupting devices can be coordinated. Proper coordination of the protective devices is paramount during a fault condition on an electrical system. The device closest to the fault needs to be opened to clear a fault from the system while an upstream device is set to allow passage of the fault current necessary for operation of the downstream device without itself tripping. Relay schemes, fuse curve coordination, and operating speed of interrupting devices play roles in this process. The separation between current characteristics curves of a coordination study at the maximum let-through current should be greater than either the operating time of the interrupting device that is to open on fault or the melting time in case of a fuse.

The advent of microprocessor-based metering and relay equipment allows for a more selective coordination of system components and also for interfacing with an industrial network so that settings and load information can be displayed at remote locations. For large motors, these devices have provided for custom protection of motors in one relay.

CONDUIT AND WIRE

The NEC defines permitted uses for each type of conduit cable support system, the installation requirements for supports, and grounding concerns (National Electrical Code, 1993). Based on NEC data, Table 6.7 catalogs the areas where different types of conduits can be used.

Table 6.7 Areas for permissible use of conduit.

Classification	Exposed	Direct burial	Hazardous areas
Rigid metal conduit	Yes	Yes	Yes
Intermediate metal conduit	Yes	Yes	Yes
Electrical metallic tubing	Yes	Yes	As noted in NEC[a]
Rigid nonmetallic conduit	Yes	Yes	As noted in NEC
Flexible metal conduit	As noted in NEC	No	As noted in NEC
Electrical nonmetallic tubing	As noted in NEC	No	No

[a] NEC = National Electrical Code.

The tables in Article 310 of the NEC, which provide the ampere capacity rating for wire and cable up to 15 000 V are based on the metal used to conduct the electricity, either aluminum or copper. Notes following the tables should be consulted for wire and cable installation derating requirements. Another source for wire and cable ampacity is IEEE Power Cable Ampacities (IPCEA Power, 1962).

The NEC provides tables for the conduit fill requirements for 600-V rated cables. For cables rated above 600 V, manufacturer's standards for cable construction will need to be consulted for cable dimensions and uses.

Guidelines for specifying electric submersible pump cable are provided in IEEE Standards 1018 (Recommended Practice for Field, 1985) and 1019 (Recommended Practice for Specifying, 1985). For wiring practice in hazardous areas, IEEE Standard 576 (Recommended Practice for Installation, 1989) and the NEC can be reviewed.

Sizing requirements of circuits for motor feeder and service conductors and the conduit systems that support them are covered in the articles of the National Electrical Code. Article 430 discusses requirements for sizing the feeder cables and disconnect for a motor circuit, and tables provide listings of motor full-load currents, which is useful when the nameplate data of the motor is not known. The listed motor ampacity is multiplied by 125% to calculate the design load for sizing feeders and disconnect for the motor, and listed motor ampacity is multiplied by 150% to calculate the full-load ampacity of the drive. The drive-calculated, full-load ampacity is then multiplied by 125% to calculate the design load for sizing feeders and disconnect for the drive.

Not covered in the NEC are sidewall pressure limitation on cables during installation, pulling tension limitations of cables during installation, and the fault-withstand rating of wires and cables (the amount of current a cable can carry over time without damage to its insulation). The cable fault withstand must be considered when sizing feeder circuits for proper coordination during fault conditions for downstream devices. Sidewall pressure and pulling tensions should be addressed in the design of the conduit system.

When designing a medium-voltage system, the designer needs to be familiar with special requirements and considerations of cables.

Cable testing is recommended to verify the reliability of the electrical system after it is installed and also as a means of periodically determining the condition of the system. Testing of cable insulation with a meggar or by dc voltage provides a basis on which to compare future cable tests.

LIGHTNING PROTECTION

Requirements for the design of a lightning protection system can be found in NFPA Standard 78 (Lightning Protection, 1989).

Lightning is a spark averaging 25 000 A and 30 000 V that discharges between the ground and a cloud in 50 microseconds across a path from a few thousand feet up to several miles. The discharge is made up of multiple pulses and generates heat in a range of 16 500°C (30 000°F) in a matter of microseconds. The heat generated by a lightning stroke can cause major damage to structures by converting moisture in building materials to high-pressure steam that can cause concrete and mortar to crack and lumber and fiberglass to shatter. The heat can also vaporize, weld and melt electrical and electronic components.

Lightning hazards include side flashes caused when adjacent objects are at different potentials because of the sudden and unequal rise in ground potential. Side flashes balance the potential between objects.

A structure's lightning protection system needs to intercept the lightning stroke, divide the current, and conduct it safely to the ground past the structure and its contents and occupants. The lightning protection system requires a low-resistance ground system that provides equal ground potential for all parts of the structure and its contents and occupants.

Another form of lightning protection is surge protection. Most lightning strokes are cloud to cloud, but this type of lightning induces a voltage to electrical and electronic components. Standards for components designed to handle induced currents are found in IEEE Standard C62-1990 (Guides and Standards, 1990).

Surge protection should be installed on telecommunication systems, security systems, fire alarm systems, motor control centers, switchboards, switchgear, computers, standby generators, surveillance television systems, instrumentation, and all systems that communicate or are connected from the facility by wire.

STANDBY POWER

Guidelines for the installation of standby power can be found in IEEE Standard 446 (Orange Book, 1987). The need and requirements for a second source of power to a pump station are dictated by the reliability of the primary power source and the requirements of local jurisdictions.

The standby source of power could be a second utility line if arrangements can be worked out with the utility serving the facility. The cost of providing a second source, which can be charged to the facility owner, may dictate installation of an on-site source of standby power (See Figure 6.2) as a more economical approach.

Engine-driven pumps are another possible secondary source of power. Space must be provided in the design for any fuel storage facility. Standby power should be provided for any fuel pumping needed to transfer fuel to the engine from the fuel storage facilities.

Figure 6.2 Standby power generator.

Standby power also can come from an engine and generator set. Sizing of the unit depends on the loads that will be applied to the unit, the sequence in which the loads will be applied and the type of loads being applied. If variable-frequency drives are included as part of the loads served by the generator, then the supplier of the generation equipment needs to be informed of these loads. Variable-frequency drives cause harmonic distortion, which can affect the generator's voltage regulator if it does not have three-phase compensation. The generator supplier may recommend an oversized generator to compensate for the effects of overheating in the generator wiring caused by harmonics distortion.

Recommendations of the engine manufacturer should be followed for intake air, exhaust, and cooling requirements. When an engine-driven pump or engine and generator set is installed, the designer must consider maintenance requirements of the engine. Space and circuits need to be provided for starting batteries, battery charger, engine heater, remote radiator, day tank, fuel level probes, and other engine ancillary equipment.

Regarding the transfer scheme in the event of power failure, the time the facility can be without power must be determined. A delay period for the transfer scheme can be specified to allow the station to ride through temporary power outages. If the generator is sized to handle only part of the loads, the transfer scheme should contain contacts to lock out loads that overload the generator. A time delay period can be specified to delay operation of the transfer scheme when operating on the standby source, and this period starts when the utility source becomes available. It maintains operation of the supply voltage from the standby source to verify that the utility source voltage is stable before the transfer takes effect.

In transferring motor loads from the standby power source back to the normal utility source, the transfer scheme can handle motor load transfer with in-phase transfer between sources, a time delay in a neutral position between the two sources during transfer, or by shutting off the motors before transfer between sources.

ADJUSTABLE-SPEED DRIVES

Alternating current, variable-frequency drives (VFDs) are the most common type of adjustable-speed drives used in wastewater and stormwater applications. They allow pump-driven ac motors to be operated at adjustable speed, resulting in adjustable flow.

The option of adjusting the speed of ac motors promotes their use over dc motors. More specifically, ac squirrel cage induction motors are simple, capable of high speed, require little maintenance, and typically perform well under such environmental conditions as dirty air and explosive atmospheres.

In the past, VFDs had numerous problems and were considered unreliable. Modern VFDs are composed of highly reliable semiconductors, which have increased their use considerably. They are now considered dependable, although still subject to several problems.

For varying torque loads such as centrifugal pumps, VFDs are capable of controlling speed over a 10:1 range. Centrifugal pump speed ranges typically are limited by the pump remaining within its operating envelope of the system head curve. As this limitation is almost always less than 10:1, VFDs rarely limit the speed range of centrifugal pumps.

TYPES OF ALTERNATING CURRENT VARIABLE-FREQUENCY DRIVES. There are three major classes of VFDs: current source, voltage-source intermediate circuit with variable link voltage (better known as voltage source), and voltage-source intermediate circuit with constant link voltage [better known as pulse-width modulation (PWM)]. Current source and PWM are the most common in wastewater and stormwater pumping applications.

The classes are defined by the type of intermediate circuit that converts constant frequency ac input power into an adjustable frequency ac output, which drives the pump motor (Dulas and Tisdale, 1992). The intermediate circuit consists of converter and invertor sections. The converter section rectifies the alternating line current into clean direct current, and the invertor section then changes the direct current back to adjustable frequency ac power to drive the pump motor.

The design decision to select current source versus PWM variable-frequency drives is not clearcut. In smaller sizes, PWM drives typically are less expensive than current source drives. At larger sizes, current source drives become cost competitive and for some sizes are cheapest. The power factor for PWM drives is better than for current source drives, so PWM energy costs typically are less. For the same current draw motor, the current source drive is physically larger than the PWM drive.

The invertor section of PWM drives is more complicated because PWM drives require high-speed switching, while the switching circuits of the current source drive generates only a six-step output, requiring lower-speed switching. Thus, PWM drives might need more maintenance.

Both type of VFDs produce harmonics, which distort input power supply and can harm nearby electronic and communication equipment. However, PWM drives require smaller filters to reduce total harmonic distortion to acceptable limits and do not produce the notches in the sine wave input power supply produced by current source VFDs. Notching can cause failure of a pumping station control system operating under emergency generator power (Keskar, 1993).

Inherent to their design, VFDs also produce harmonics in their output power to the driven motor. Harmonics generate heat, thus VFD-driven motors run hotter than constant-speed motors under the same conditions. Because PWM drives produce more harmonics than current-source drives, it is recommended that a motor driven by a PWM drive be rated for inverter duty service. Inverter duty service motors are furnished with Class F insulation and rated for a "B" temperature rise. The conservative choice is to specify inverter duty service motors for current source drives as well.

Matching Variable-frequency Drives and Motors. Variable-frequency drives systems require proper matching of the motor and drive, and mismatches sometimes occur (Dulas and Tisdale, 1992). Torque requirements at any point in the operating speed range play a role, and a common mistake is to use the nameplate power rating to match drive and motor. Because different motor types have different power factors, full load amperage draw may differ even though the power rating is the same. Full load amperage draw should be used in matching motor and drive (Keskar, 1993).

Several methods have been used in the design documents to promote the matching of motor and drive. In one, the pump supplier is assigned complete responsibility for the match and purchases the VFD from a vendor. A drawback with this approach is the possibility of projects having different types of pumps supplied by multiple pump suppliers. Obtaining VFDs of only one manufacturer for the entire project can prove difficult. For stations where one type of pump is specified, this is less of a problem.

In another method, the electrical subcontractor supplies the VFDs for the entire project and mounts them together with the motor control centers for constant-speed pumps. The electrical subcontractor assumes some responsibility for the match but does not purchase the pumps. This promotes installations that have only one VFD manufacturer and normally makes for a well-coordinated motor control center. However, because the electrical subcontractor does not purchase the pumps, the responsibility of correcting the problem is less clear than in the first method if a mismatch occurs.

Variable-frequency Drive Interfaces. Variable-frequency drives are required to interface with the electrical system and the instrumentation and control system. Proper coordination of control circuits and speed signals with these systems is essential.

It is recommended that control circuits be 115 V–ac, which is standard or an available option on most drives other than some smaller VFDs. For these VFDs, the control circuit is low ac or dc voltage, for example 12 V–dc. Low-voltage control circuits are prone to outside interference such as from voltage induced from nearby 460-V, three-phase ac power.

Often, VFDs are controlled by an external analog follower signal, which must be compatible with the process control system. This signal typically will be 4- to 20-mA speed input as opposed to a varying voltage signal such as 1- to 5-V–dc. Similarly, the speed output signal must be compatible with the process control system and also typically will be 4 to 20 mA, isolated if possible. The standard interface signals are varying voltage such as 1 to 5 V–dc, especially for some smaller VFDs. Thus, if not properly specified, field problems can result.

Some Problems with Variable-frequency Drives. Many VFDs are rated for service between 0 and 40°C (32 and 104°F) with a noncondensing relative humidity of up to 95%. They typically are installed in a non-air-conditioned electrical room. In years past, the electrical room was air conditioned to provide the VFD solid-state components a mild environment, and this practice may still be appropriate for selected applications.

If a bubbler tube is used for level control, avoid mounting the bubbler tube pressure transmitter in the VFD enclosure or it might introduce corrosive gases such as hydrogen sulfide, damaging the VFD electronics.

The solution to harmonics and notching problems is to reduce the total harmonic distortion of the incoming power supply. Typical solutions include adding filters and reactor circuits to the VFD. Reducing harmonics involves not only the drive itself, but how it interacts with the incoming power, and some practitioners recommend that design documents emphasize this interaction (Keskar, 1993). For instance, the VFD drive vendor might receive more information on the electrical system design than is now commonly issued.

The applicable standard for harmonic control is IEEE Standard 519 (Guide for Harmonic, 1981), which places limits on such items as total harmonic distortion.

REFERENCES

Circuit Breakers, Switchgears, Relays, Substations, and Fuses (1989). Stand. C37, Inst. of Electr. and Electronics Eng., Inc., New York, N.Y.

Distribution, Power, and Regulating Transformers (1990). Stand. C57, Inst. of Electr. and Electronics Eng., Inc., New York, N.Y.

Dulas, J.H., and Tisdale, J.C. (1992) Improve Flow Control with Electronic Drives. *Chem. Eng. Prog.*

Gold Book (1990). Stand. 493, Inst. of Electr. and Electronics Eng., Inc., New York, N.Y.

Green Book (1982). Stand. 142, Inst. of Electr. and Electronics Eng., Inc., New York, N.Y.

Guide for Safety in AC Substation: Grounding (1986). Stand. 80, Inst. of Electr. and Electronics Eng., Inc., New York, N.Y.

Guide for the Interconnection of User-Owned Substations to Electrical Utilities (1990). Stand. 1109, Inst. of Electr. and Electronics Eng., Inc., New York, N.Y.

Guides and Standards for Surge Protection (1990). Stand. C62, Inst. of Electr. and Electronics Eng., Inc., New York, N.Y.

Guide for Harmonic Control and Reactive Compensation of Static Power Converters (1981). Stand. 519, Inst. of Electr. and Electronics Eng., Inc., New York, N.Y.

IPCEA Power Cables Ampacities (1962). Inst. of Electr. and Electronics Eng., Inc., New York, N.Y.

Keskar, P.Y., CH2M Hill, Gainesville, Fla. (1993) Personal Communication.

Lightning Protection Code (1989). Stand. 78, Natl. Fire Protection Assoc., Quincy, Mass.

National Electrical Code (1993). Stand. 70, Natl. Fire Protection Assoc., Quincy, Mass.

National Electrical Safety Code (1990). Stand. C2, Inst. of Electr. and Electronics Eng., Inc., New York, N.Y.

Orange Book—Recommended Practice for Emergency and Standby Power for Industrial and Commercial Power Systems (1987). Stand. 446, Inst. of Electr. and Electronics Eng., Inc., New York, N.Y.

Recommended Practice for Field Testing Electrical Submersible Pump Cable (1985). Stand. 1018, Inst. of Electr. and Electronics Eng., Inc., New York, N.Y.

Recommended Practice for Fire Protection in Wastewater Treatment Plants (1990). Stand. 820, Natl. Fire Protection Assoc., Quincy, Mass.

Recommended Practice for Installation, Termination, and Testing of Insulated Power Cables Used in the Petroleum and Chemical Industry (1989). Stand. 576, Inst. of Electr. and Electronics Eng., Inc., New York, N.Y.

Recommended Practice for Specifying Electrical Submersible Pump Cable-Ethylene-Propylene Rubber Insulation (1985). Stand. 1019, Inst. of Electr. and Electronics Eng., Inc., New York, N.Y.

Red Book—Recommended Practice for Electrical Power Distribution for Industrial Plants (1986). Stand. 141, Inst. of Electr. and Electronics Eng., Inc., New York, N.Y.

U.S. EPA (1974) Technical Design Criteria for Mechanical, Electrical, and Fluid System Component Reliability. EPA-430/99-74-001, Washington, D.C.

Chapter 7
Instrumentation and Control Systems

Control technology for wastewater and stormwater pumping stations has advanced greatly over the last decade, mainly as a result of the microprocessor, which now pervades control applications. Programmable controllers, digital telemetry systems, and a new generation of transmitters are all microprocessor based. To the pumping station designer, these advances offer increased reliability, improved ways to meet design objectives and alternate equipment, a reduction in labor, and valuable data.

Relevant design objectives include proper functioning of the pumping station and proper management of the overall wastewater or stormwater pumping system. Improved ways of alternating equipment can extend equipment life, and a reliable pumping system reduces overflows and surcharges. A reliable telemetry system often reduces labor and operation costs. Data from catchment areas can identify infiltration and inflow problems. Flow data assists in designing for expansions. Data characterizing diurnal patterns can identify frequently operating equipment, which can be incorporated into a maintenance program.

PROCESS AND INSTRUMENTATION DIAGRAMS

The use of process and instrumentation diagrams (P&IDs) and process flow diagrams is suggested in the design of wastewater and stormwater pumping stations because they provide an efficient means of communicating design concepts. Use of P&IDs is recommended even for simple stations, although the designer may choose not to include them as part of the design documents.

An example P&ID of a basic pumping station is shown in Figure 7.1. Symbols of P&IDs should conform to Instrument Society of America Standards and Practices, specifically S5.1—Instrumentation Symbols and Identification (ISA, 1992).

Within their simplicity lies the value of P&IDs. The design of a pumping station is multidisciplinary and P&IDs promote communication between process mechanical, instrumentation and control (I&C), and electrical designers. They promote proper piping and process concepts and the creation of an accurate instrument list and assist in establishing control concepts and subsequent development of control diagrams. Accurate electrical drawings with detailed conduit and conductor runs have as their origin correct P&IDs. With P&IDs, pumping station operations personnel can input their ideas to the design.

During design, P&IDs are not static and reflect the changes brought about by the design team itself, quality assurance reviewers, client preferences, and value engineering efforts. However, the design effort is usually improved if the P&IDs are "frozen" early in the design, with only minor changes made as the design progresses.

LEVEL MEASUREMENT

Early in the design, the required operating level range should be determined because it impacts vertical dimension of the wet well and hence the structural design. Often overlooked is the fact that enough vertical dimension should be reserved to sense abnormal low and high levels (besides normal high and low operating points). Also suggested is to sense these abnormal levels by primary elements independent of the elements used to sense normal operating levels.

The abnormal low level should be at least equal to the center line elevation of the impeller. The conservative choice is to locate the abnormal low point equal to the volute top, to reduce air entry into the volute. For the case of submersible pumps, the submersible motor may add another complication. The conservative choice is to locate the abnormal low point above the top of the submersible motor for cooling, increasing the required vertical dimension even more. However, most submersible pumps can operate properly with their motors only partially submerged. If the designer wants to reduce the vertical dimension or operate the submersible pump at times with its motor nonsubmerged, the pump manufacturer should be contacted for a recommendation.

The normal low operating level is typically set at least 100 to 150 mm (4 to 6 in.) above the abnormal operating low level, though the normal low operating level should be confirmed with the pump vendor. Based on wet well geometry and piping layout, vendors sometimes recommend a normal low

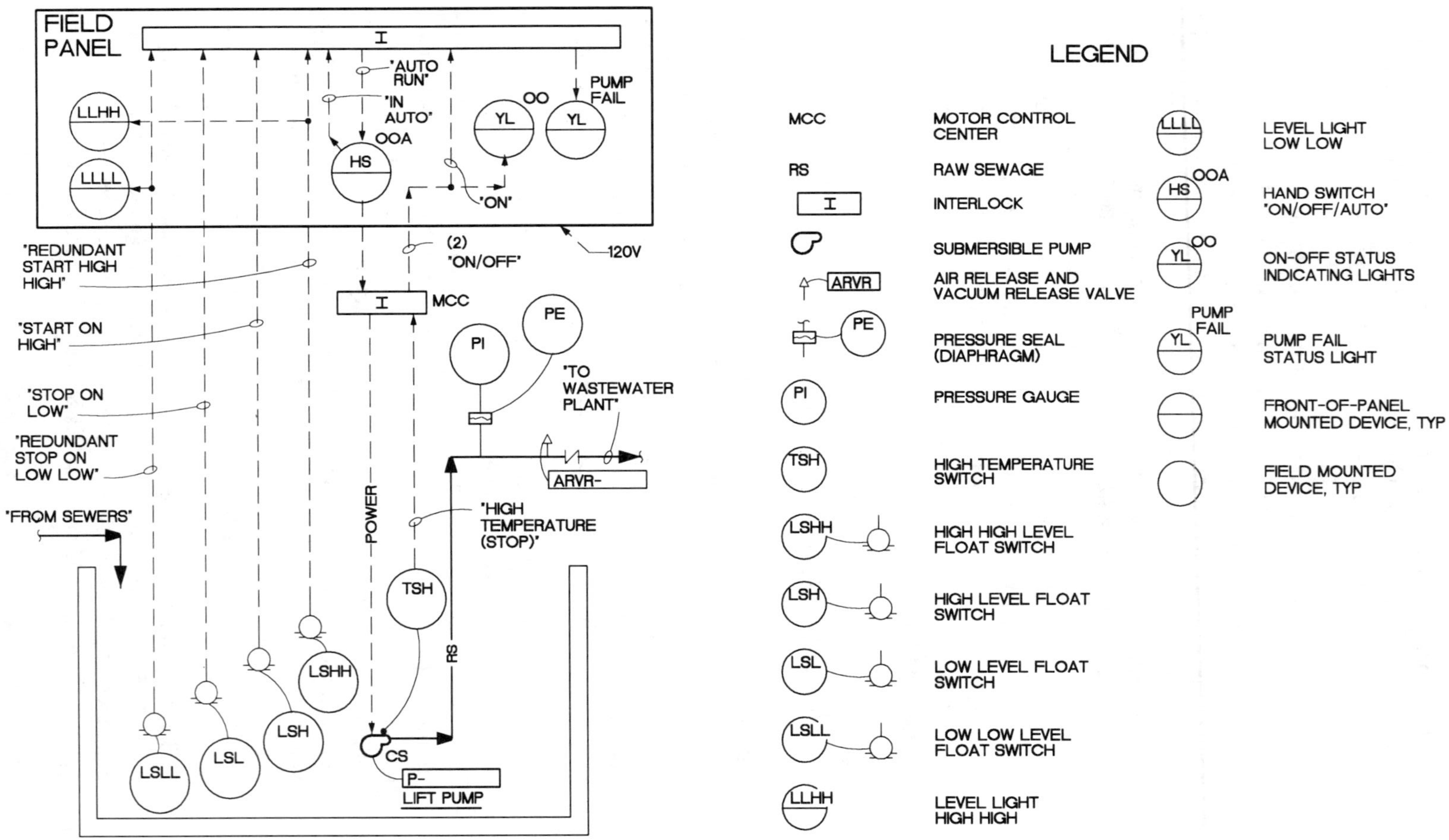

Figure 7.1 Example process and instrumentation diagram of a basic pumping station.

operating level to ensure proper operation, for example, reducing vortexing effects to meet minimum net positive suction head (NPSH) requirements.

Wet well level sensing elements typically used include float switches, pneumatic bubbler systems, ultrasonic sensors, and submersible level sensors.

FLOAT SWITCHES. Although capable of only sensing fixed setpoints, float switches (floats) typically are the most cost-effective level element. To resist chemical action, floats should be made of plastic or other resistant material. They should be put in a quiescent area away from the inflow point. Floats are subject to jamming or damage by solids and fouling from accumulations of grease, fats, oil, scum and debris. Periodic cleansing is required, and for certain applications, continuous flushing may be needed.

Floats can be mounted either to a vertical pipe [typically 25 mm (1 in.) in diameter] or to a suspended cable furnished with a weight at its bottom. Vertical pipe mounting is suggested for most applications because the setpoints are reproducible and it requires less space in plan view than suspended cable mounting, which requires more area because the float can move 360 degrees in plan view on a suspended cable. Also, debris can cause the suspended cable to rise, resulting in malfunction. Suspended cables do allow the changing of setpoints to be made easier than with the vertical pipe. With either mounting, float setpoints may be as close as 100 mm (4 in.) apart.

To prevent tangling of close setpoints, floats should be staggered on each vertical pipe or mounted on different pipes. In a staggered arrangement, in plan view, the floats are not mounted on top of each other, but rather form angles with each other.

As with all primary elements, level elements must be approved for use in their surrounding atmosphere. Specific care should be given to floats because they are located in the wet well itself. Wastewater wet wells are often classified as Class I, Division 1 hazardous locations. Accordingly, float circuits must be provided with intrinsically safe relays.

If the wet well is classified as hazardous, then conduits and conductors must meet required codes for use in hazardous atmospheres. See Chapters 3 and 6 for more details.

BUBBLER SYSTEMS. The pneumatic bubbler system can either monitor discrete level setpoints with pressure switches or continuously measure wet well level with a pressure transmitter. The bubbler system also requires a compressor, receiver, ancillaries, and a bubbler tube immersed in the wet well. Ancillaries include such items as a pressure-regulating filter, constant differential pressure relay, a blowdown valve, and isolation valves.

Except for the bubbler tube itself, these components are available in an integral control panel, useful in isolated pumping stations. Where the pumping station is adjacent to a complex such as a wastewater plant, the bubbler system is more custom designed, often making use of existing high-pressure air systems. The receiver typically is furnished with a low-pressure switch, which monitors compressor malfunction. As with floats, the tube is subject to fouling. To clear the tube, provisions typically are made to subject it to a high-pressure air purge, which can be done manually or automatically at periodic intervals. Sometimes provisions are made to physically rod the tube.

Bubblers generally are reliable and cost effective. However, especially because of the number of fittings, some systems are subject to leaks, leading to erratic results. During cold winter months, some systems perform erratically because of freezing of condensed moisture within the tubing.

In corrosive environments, the working life of bubbler systems has been reduced by corrosion of system components. Accordingly, component materials should be as noncorrosive as possible. Bubbler systems located indoors can be installed in the same room as the electrical switchgear, though corrosion problems with electrical equipment have been experienced in some locations because of corrosive gases introduced by bubbler tube leaks. This has led some practitioners to avoid locating bubbler systems with electrical switchgear.

A bubbler tube 13 mm (0.5 in.) in diameter with its end approximately 150 mm (6 in.) from the wet well bottom has proven adequate. Some commercially available bubblers deliver an air flow of approximately 0.2 L/s (0.4 scfm), though some practitioners prefer a higher flow of approximately 0.5 L/s (1 scfm) (Bergs, 1993).

Pressure switch designs should be reviewed critically, as the commercially available ranges of pressure switches are limited. In certain wastewater and stormwater applications, suitable pressure switches are only available in the low end of the ranges. Switch deadband is sometimes large enough at some low ends to render the switch nonreproducible and useless (Whitenack, 1992).

Figure 7.2 shows a typical bubbler system within a control panel. This system happens to use pressure switches, though the compressor is not integral to the panel. Note the high-pressure purge bypass around the normal operation piping. Figure 7.3 depicts the top of a bubbler tube furnished with a plug, which can be removed to allow rodding the tube.

ULTRASONIC SENSORS. The use of ultrasonic level sensors is increasing because of improved electronics and their nonintrusive nature, which makes them less subject to fouling and debris. Particularly in stormwater pumping applications, they are often the element of choice because they are typically not affected by debris.

On occasion, ultrasonic level sensors are affected by large floating materials. They are confused by foam, vortices, turbulence and wall echoes. Where the ultrasonic transducer does not include its own reference target for temperature or humidity correction, thermistor temperature monitoring and measurement correction have been incorporated to some transducer designs. The thermistors can fail and render the transit time correction algorithms

Figure 7.2 Bubbler system internals.

ineffective. While the sensor is nonintrusive (not immersed in the wet well liquid), it must still be mounted in the well, exposed to the ambient atmosphere.

The sensor must be mounted far enough away from walls to prevent wall echo interference. Mounting should also allow sufficient headroom for the ultrasonic emission deadzone [typically 460 to 610 mm (18 to 24 in.)] above the absolute maximum liquid high level]. Sometimes wet well bottoms are grouted to form sharp angles to prevent solids and debris from gathering in

Figure 7.3 Bubbler tube top.

corners. If not accounted for, such grouted surfaces may interfere with the desired range of the sensor. A wastewater wet well with a closed access hatch and no grating can be considered a closed tank. Some but not all ultrasonic level meters require that the bottom of their megaphones be at least 50 mm (2 in.) above the vent pipe of the wet well to ensure safe operation if the wet well is classified as Class I, Division 1 and is considered a closed tank.

SUBMERSIBLE LEVEL SENSORS. Submersible level sensors are submerged near the bottom of the wet well, and measure level by sensing hydrostatic head pressure.

The sensor itself is a polytetrafluoroethylene-faced diaphragm connected by cables to a transmitter housing. The diaphragm is part of a sealed, oil-filled system. A variable-capacitance transducer converts the pressure change to an electronic signal. As with any pressure-based system that measures level, specific gravity of the measured fluid must remain constant, which is essentially the case for wastewater and stormwater.

A sealed breathing system is a key part of the system, and can accommodate cable lengths of up to approximately 24 m (80 ft). The breathing system protects the internals of the submerged sensor, allows for expansion and contraction, and nulls the effects of barometric changes.

These units, available either loop powered or externally powered, are not intrinsically safe, but can be made so by the use of intrinsically safe barriers and other arrangements.

The popularity of these devices is increasing, and practitioners generally have been pleased with this type of system. Because they are submersed, they have particular value where freezing is a concern.

These units can be mounted either on a vertical pipe or by means of a suspended cable. As with float switches, the conservative choice is to pipe mount.

FLOW MEASUREMENT

Flow meters are installed on most pumping stations. In smaller collection systems, flow meters might not be used. Flow meters typically will be installed on larger pumping stations that discharge directly to wastewater treatment plants. For wastewater or stormwater systems in which flow is managed systemwide, flow meters are prevalent.

An accurate flow meter facilitates pump field tests, which are difficult to perform precisely even under the best of circumstances. The acquired flow data also assists in systemwide planning for future expansion. Instantaneous flow data aid in the operation of wastewater or stormwater systems where diversion and storage are practiced.

Flow can be measured either in a closed pipe or an open channel. Closed pipe flow measurement is more common in wastewater and stormwater pumping stations. Open channel flow measurement typically is reserved for those pumping stations with screening facilities that require an open channel. Commonly used closed pipe flow meters include magnetic flow meters, ultrasonic flow meters, and differential producers. Open channel flow measuring systems include flumes and nozzles and the "secondaries" that measure their levels and convert them to flow. Commonly used "secondaries" include ultrasonic and admittance sensors.

All closed pipe flow meters will not measure accurately if a pipe is partially full, which can occur in horizontal runs under low or no flow conditions. If such flow conditions are expected, make provisions to reject the flow measurements by monitoring the process fluid level in the pipe. This "full/not full" switch can be used to discriminate and select flow information obtained only during suitable measuring periods.

Early in the design, inline meters should be selected and sized. Such inline meters include magnetic flow meters, flow tubes, and classic Venturi meters. By themselves, these meters typically do not produce a significant (nonrecovered) head loss. However, to measure accurately, these meters may need to be less than line size, and the fittings required can produce a significant head loss. Although more of a problem with gravity systems, this head loss can impact pump selection.

For both open and closed channel flow meters, a certain number of straight upstream and downstream diameters are required for accurate flow measurement. This number depends on both the type of meter and the type of upstream or downstream obstruction. Generally, upstream effects are more pronounced, requiring longer straight upstream runs than downstream runs.

There are many pumping stations in which flow is consistently measured accurately, but also many stations where flow measurement is inaccurate. Accurate flow measurement probably is less common than accurate level or pressure measurement. Besides following guidelines found in this and other books, manufacturer's recommendations found in their operations manuals and metering guides should be followed.

The design objective of flow measurement is to provide accurate and reproducible data. Occasionally, only reasonably reproducible and not overly accurate flow data are required. Thus, recommended flow meter installation practices sometimes are violated. A common example is installing flow meters to existing piping at less than recommended straight upstream diameters. In these instances, owners should be apprised of risks such as unacceptable reproducibility before installing flow meters.

Published accuracy of flow meters usually is done as either a percentage of "rate" (accuracy as compared to the actual flow rate) or a percentage of "span" (accuracy as compared to the full-range flow rate). Percent of "rate" values are clearcut. Percent of "span" values are less so, as there may be unacceptable inaccuracies at low end flows.

MAGNETIC FLOW METERS. Magnetic flow meters are well suited for stormwater and wastewater applications because they are chemically resistant if properly specified and are little affected by sediments and debris. Their accuracy typically is ±1% of rate over a flow range of 10:1. Suitable liner material for wastewater is either polyurethane or polytetrafluoroethylene. In most cases, polyurethane is as suitable as polytetrafluoroethylene and is the liner of choice because it is less costly. Though 316 stainless steel electrodes typically are used, there are other satisfactory electrode materials.

There are two basic types of magnetic flow meters, ac and pulsed dc, which refer to how the magnetic field of the meter is excited. The pulsed dc type is predominant in wastewater and stormwater pumping applications. Although its field is excited by pulsed dc current, supply power for this type of meter typically is 120 V–ac.

Magnetic flow meters typically require at least five straight upstream and two or three straight downstream pipe diameters. Models are available with even shorter requirements.

Exceptions to the five straight upstream diameter rule are the distances from control valves and from pump discharge, for which the meter may require 10 straight upstream pipe diameters. The 10 straight upstream rule is necessary to ensure high accuracy (for example, 0.5% of rate, of which some modern microprocessor-based magnetic flow meters are capable). Often, 10 straight upstream pipe diameter is impractical to achieve and high accuracy is not needed in stormwater and wastewater applications. The meter manufacturer should be contacted when a shorter straight upstream run is contemplated.

Where magnetic flow meters are less than line size, some manufacturers permit concentric reducers to be placed immediately on both sides of the meter.

Magnetic flow meters are available with transmitters mounted either on the meter body itself or remotely from it. Remote-mounted transmitters typically are located near the meter, though they can be located farther away. Remote-mounted transmitters reduce possible damage from hydraulic transients and slamming check valves.

Besides watertight ratings, magnetic flow meters are available in options for accidental (up to 48 hours) and even continuous submergence use [both of up to approximately 9 m (30 ft)]. Some manufacturers also place an upper time limit on their continuous-submergence magnetic flow meters. The designer should be conservative in option selection, because moisture in the electronics can render the meter useless.

Magnetic flow meters have several standard or optional features that may be of benefit. Some have a zero return feature that can be interlocked with either a pump or a device measuring pipe level. If the pumps stop or if a pipe is partially empty, the meter reading will be zero. Low flow cutoff set typically at 1 or 2% will suspend meter operation. Sloshing within the pipe at actual zero flow may produce a false flow reading but for this switch. As an option, magnetic flow meters are capable of reading bidirectional flow, which might be of use in some applications.

Entrapped air will produce erroneously high readings. In many pumping stations, the horizontal line in which the magnetic flow meter is placed is a local high spot. Such installations are potential sources of error because of entrapped air, and manual and automatic air release valves should be used to release air. Preferred practice is to lower the meter below the horizontal line, or to slope the horizontal line piping so downstream piping is raised enough to guarantee a full magnetic flow meter.

Improper grounding is a common source of erratic magnetic flow meter readings. Even where adjacent pipe is conductive and serves as a ground, bonding jumpers can corrode.

Although used in the past, the trend is now *not* to furnish ultrasonic power generators to clean electrodes (Cooper, 1992). Ultrasonic cleaning is more appropriate for crystalline substances than for greases and biological growths found in wastewater pumping stations. However, the designer can select a magnetic flow meter, which can be retrofitted with either a permanent or portable ultrasonic power generator.

Compared to ultrasonic meters, magnetic flow meters are more expensive in larger line sizes. They are inserted directly to and actually substitute for the pipeline, compared to many ultrasonic models that clamp to the outside of the pipe.

Frequently, magnetic flow meter size is smaller than the connecting process pipe because the meter is designed with a minimum velocity to keep debris suspended and reduce electrode coating. As a guideline, this minimum velocity typically might be 0.6 to 0.9 m/s (2 to 3 ft/sec) for stormwater and raw wastewater applications (deGlanville, 1992). Under this guideline, for 10:1 flow ranges, maximum flows produce line velocities of 6 to 9 m/s (20 to 30 ft/sec), which may be hydraulically unacceptable. An alternate guideline is that minimum velocities of 0.15 to 0.3 m/s (0.5 to 1.0 ft/sec) are acceptable, provided flushing velocities of at least 1.8 m/s (6 ft/sec) are present daily (Gunby, 1993).

Providing bypass piping around a magnetic flow meter should be considered, especially in a pumping station where downtime cannot be tolerated. If a bypass is used, an access flange to allow cleaning the inside of the meter should be evaluated. In noncritical stations, the magnetic flow meter taken out of service might be replaced with a spool piece.

Field replaceable electrodes are not widely used in magnetic flow meters measuring stormwater and wastewater flow, but may be beneficial in some applications. Electrodes have been known to develop coats and lose accuracy in some uses, for example with thick sludges. For stormwater and wastewater applications, comparable coating may take years to occur, if ever. However, certain industrial wastewaters can foul the electrode quickly. Thus, especially for large meters with routinely low flow rates, field replaceable electrodes may be worth the extra expense in certain applications.

For the same reason, "electrodeless" magnetic flow meters might also be beneficial (Gemin, 1993).

The designer should be aware of the term *field replaceable*, which is used somewhat ambiguously. Most operations personnel desire removing the electrode from outside the magnetic flow meter, preferably under process conditions. If so, a field replaceable electrode with a hot tap assembly is needed. By contrast, some field replaceable electrodes require that the electrode be removed from inside the meter, which involves shutting down the process and expending labor.

Figure 7.4 depicts a magnetic flow meter schematic. Figure 7.5 shows a horizontal installation as recommended by one manufacturer. Note that a full magnetic flow meter is ensured. Figure 7.6 shows a vertical installation. Note the optional access flange for cleaning. Field-installed meters may have a solar shield and a surge suppressor protecting the transmitter.

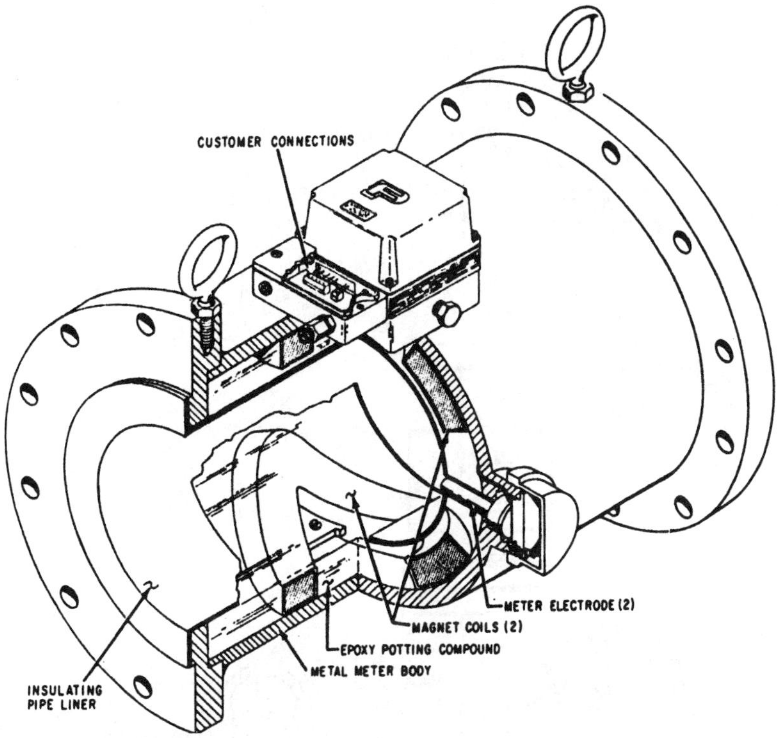

Figure 7.4 Magnetic flow meter schematic.

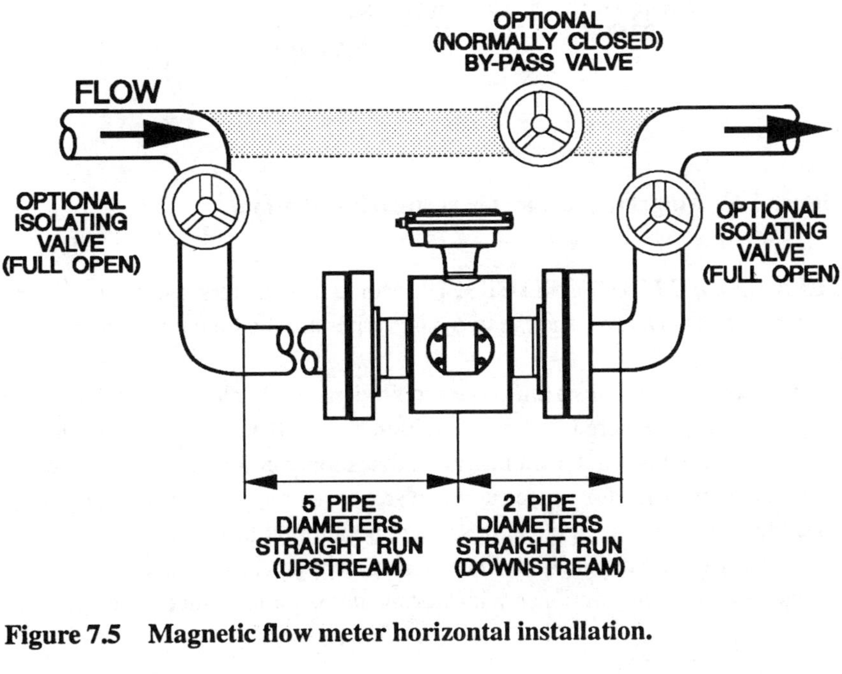

Figure 7.5 Magnetic flow meter horizontal installation.

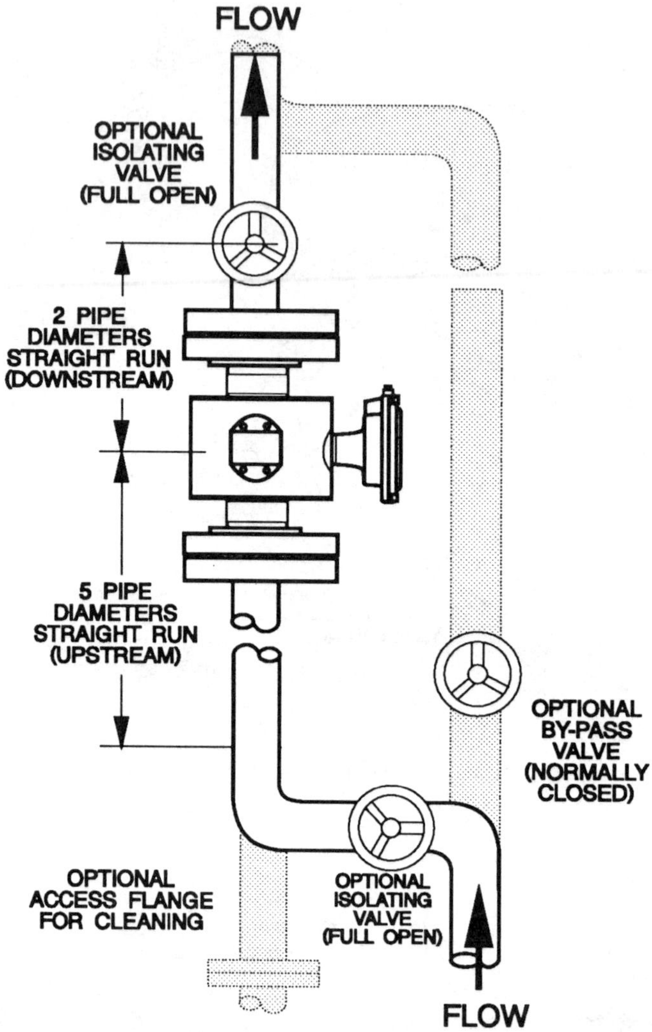

Figure 7.6 Magnetic flow meter vertical installation.

ULTRASONIC FLOW METERS. Ultrasonic flow meters also are used for wastewater applications, and the two basic types are the transit time and the Doppler.

Traditionally, the transit time type required primarily clean fluids, while the Doppler type required particles, bubbles, or turbulence. Because of microprocessor advances, some manufacturers that supply both meters now recommend the transit time for wastewater and stormwater applications. In any case, there is overlap in applications where both might be used. Wastewater is a common Doppler application. The transit time type is used for wastewater and stormwater, but experience with stormwater is more limited. The specific application must be discussed with the transit time meter manufacturer, because stormwater is not necessarily clean during the first flush and may not be clean for a long time.

Ultrasonic meter accuracy typically is advertised as ±2% of full scale. Some practitioners agree with this value (Gemin, 1993), but others (Guagno, 1992) put the accuracy at approximately 5 to 8% of rate. Range is not explicitly stated in much of the manufacturers' literature, in part because it changes as the meter's span is adjusted. Straight upstream and downstream requirements are ideally 10 to 20 and five pipe diameters, respectively. But some practitioners have had satisfactory performance by following the same laying lengths as recommended for the classic Venturi (Gemin, 1993). Minimum measurable velocity typically is 0.15 m/s (0.5 ft/sec).

A main advantage of ultrasonic meters is that in many models the transducer either clamps on or is track mounted to the pipe exterior, thus is nonintrusive. Pipe material must be able to conduct sound. Lined pipes may work, depending on the meter manufacturer and the extent of bonding. The conservative design choice is to provide an unlined spool piece for subsequent transducer installation. Alternatively, some transit time designs include a line-size spool piece in which the transducer is integrally mounted.

From an installed cost standpoint, ultrasonic meters are attractive for larger line sizes because transducer size is independent of line size. Although ultrasonic meters have been used successfully in wastewater applications, success has not been uniform (Robinson, 1986). In part, this is because of variations in flow profile, liquid sonic conductivity, and the amount of air bubbles or particles, which affect reflectivity.

The user should fully explore the specific application with the meter manufacturer, which might include field testing. Other design considerations are to reduce pipe vibration, mount the transducer on the side of horizontal pipes to reduce foaming and sedimentation interference, and mount the transducer head away from a noise source such as a pump.

The transducer is mounted to the outside of a clean, bare metal pipe by means of a mounting track or clamps, coupling grease and metal straps. A common problem is the transducer losing its bonding to the pipe (Ouzts, 1992).

Figure 7.7 shows a transit time ultrasonic flow meter, including sensors and transmitter. The top sensor is a direct mode type, while the lower sensor is a reflect mode type. Figure 7.8 shows a field installed Doppler ultrasonic flow meter. Only the sensor (not the transmitter) is shown. Note how the sensor clamps on to the pipe exterior.

DIFFERENTIAL PRODUCERS. Differential producers such as insert flow tubes, classic Venturi tubes, and orifice plates represent another class of meters used in stormwater and wastewater applications.

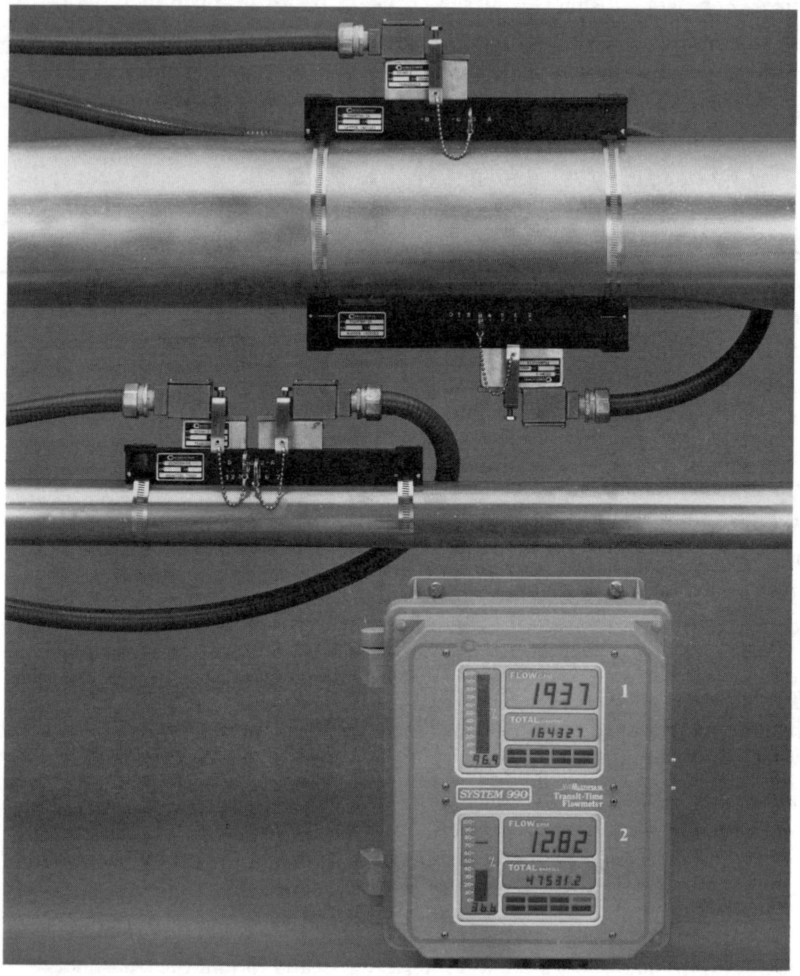

Figure 7.7 Transit time ultrasonic flow meter.

Among practitioners, there appears to be a wide difference of opinion on the use of differential producers in wastewater applications. In the opinion of some, their use should be avoided altogether because of clogging problems of the transmitter connecting tubing, capillary system problems and the concomitant increase in maintenance. Other practitioners have had successful experience and unequivocally recommend them.

Accordingly, the practitioner considering differential producers should carefully evaluate them. Manufacturers should be contacted directly for their recommendations. Information on existing wastewater installations should be gathered to determine which hardware and installation practices are successful.

The most common differential producers used in wastewater and stormwater are the classic Venturi tube and the flow tube, an insert version of the

Figure 7.8 Doppler ultrasonic flow meter.

classic Venturi. The least expensive differential producer is the orifice plate, but its use is limited in wastewater and stormwater applications. It is energy inefficient because most of its differential pressure is nonrecoverable and it is more susceptible to debris collection.

Some classic Venturi meters can measure bidirectional flow, which may be of use in some applications. Because it is less costly, the flow tube is probably more often used than the classic Venturi. Although the classic Venturi usually has a circular cross-sectional area, rectangular cross section Venturi meters are available to mate with rectangular concrete pipe sometimes found in larger pumping stations.

Differential producers share the disadvantage that their pressure taps and connecting tubing between the meter itself and the electronic transmitter tend to become clogged from debris and solids. To combat this, water purges and sealed capillary systems are used.

Water purges can be either continuous or intermittent. Continuous purges decrease system accuracy by approximately 1% of rate because the purge water adds backpressure to the differential pressure measurement (Briggs, 1993). Intermittent purges have been used successfully (Ouzts, 1992). Intermittent purges can be initiated either manually or automatically at periodic time intervals. However, intermittent purges produce intermittent "blips" in the flow data. This typically is not a concern in wastewater and stormwater.

In sealed capillary systems, the connecting tubing is fluid filled and isolated from the process fluid by the seal. In the past, users of sealed capillary systems have experienced problems of inaccuracy because of differential temperature heating of the two capillaries, increased inaccuracy at low flows, and calibration difficulties. To alleviate these problems, capillary tubes should be bundled together to reduce temperature gradients. Some manufacturers are now offering modified seal designs. More convenient calibration techniques also are becoming available.

Overall system accuracy for different producers is in the range of ±0.25 to 5.0% of rate depending on meter type, flow turndown, bore calculation errors, errors because of the capillary system (if used), and errors of the differential pressure transmitter.

Because flow varies as the square of differential pressure, the accuracy of differential pressure measurement limits flow turndown to only approximately 3.5:1 (newer, more accurate designs such as "smart" transmitters increase the turndown). Such low turndown promotes designs in which a flow meter is dedicated to each pump, as individual pump turndown is less than the overall turndown of the pumping station.

To increase turndown, sometimes two differential pressure transmitters with overlapping ranges are connected to one meter. An attendant electronics unit switches to the transmitter of the appropriate range. A proprietary unit that accomplishes this switching is available.

The coefficient of discharge for a differential producer changes below a certain Reynolds number (approximately 70 000 to 80 000), and inaccuracy will result. Accordingly, the Reynolds number should be checked, particularly at low flows.

Most of the differential produced by the classic Venturi or flow tube is recoverable, unlike an orifice plate which has much greater nonrecoverable losses. Typically, the nonrecovered head in the classic Venturi or flow tube is approximately 5%. For water applications, a typical, recoverable head loss at maximum flow for the classic Venturi or flow tube is approximately 25 kPa (100 in. H_2O). Thus, the nonrecovered or lost differential is only approximately 1.2 kPa (5 in. H_2O).

Usually the head permanently lost by the classic Venturi or flow tube by itself is inconsequential. However, these tubes sometimes need to be downsized to develop the required differential pressure for accurate measurement. Although more of a concern in gravity systems, the fittings required for the downsizing create a loss which can be substantial to the point of affecting pump selection. Accordingly, Venturi or flow tube selection should occur early in the design so it can be incorporated to pump selection.

Throats of differential producers used in stormwater and wastewater applications can be either bronze or stainless steel. Stainless steel is the conservative design choice. Throat dimensions are integral to producing accurate data. Over time, stainless steel throats might maintain their dimensions more than bronze, which is softer and may be more prone to galvanic corrosion because it is more dissimilar to the ferric tube or pipe. Stainless steel is attacked by chlorine and, if chlorine is expected, the application should be reviewed with the manufacturer.

Regardless of the differential producer used, the differential-measuring transmitter should be located below the pressure taps and the connecting tubing should be gently sloped. Entrapped air bubbles will thus be reduced, although lines should still be bled periodically. In cold weather areas, connecting tubing located outdoors should be both heat traced and insulated.

The required straight run of upstream pipe can be substantial and depends on the type of differential meter, the beta ratio (diameter of throat to inside pipe diameter) and the type of flow disturbance (for example, elbow or valve). A classic Venturi tube of β 0.6 requires seven straight upstream diameters if the upstream obstruction is a reducer. If the obstruction is two elbows not in the same plane, the required straight upstream length increases to 16.

For isolation valves upstream of the differential producers, some manufacturers specifically recommend against the use of butterfly valves (DiNapoli, 1993). In its fully open position, the butterfly valve is in the middle of the flowstream and introduces hydraulic disturbances that might affect the meter's performance.

Major manufacturers of differential producers have their own tables and curves, which recommend required lengths of straight upstream and downstream diameters for various hydraulic disturbances. These recommendations are based on actual test data. Some differential producers are fabricated by minor manufacturers that do not perform similar tests. Obtaining meters without supporting test data increases the risk of poor flow measurement.

A proprietary differential producer exists whose typical applications have been for wastewater sludges, but this device may not be applicable for wastewater or stormwater flow. A spool of concentric circular cross section is cast with a wedge-shaped "dent" on the top half that induces a pressure differential across the wedge. Wafers with pressure taps are cast into the spool on either side of the wedge. A closed capillary system transfers the differential pressure to a transmitter. Flow path is straight through the bottom of the spool, so no solids settle-out or accumulate. Pressure recovery downstream is good.

Figure 7.9 shows a continuous purge assembly for a Venturi meter.

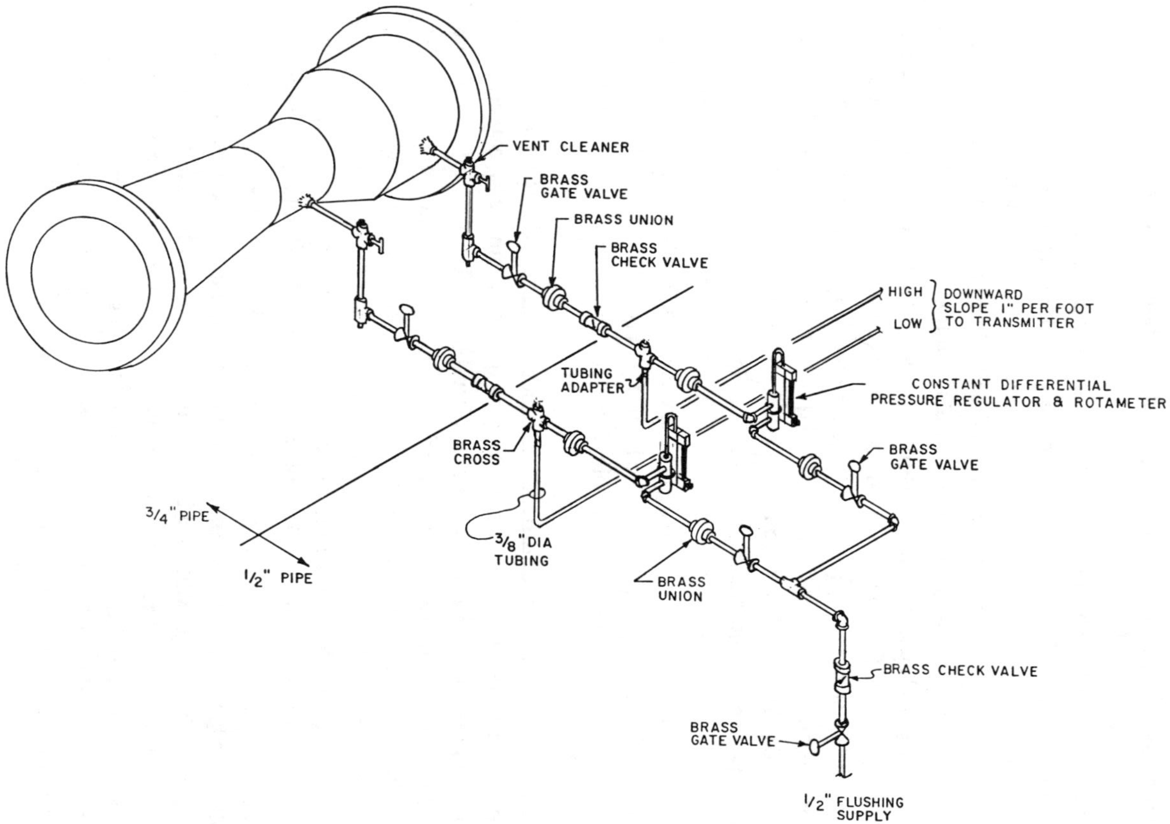

Figure 7.9 **Continuous purge assembly for a Venturi meter (ft × 0.304 8 = m and in. × 25.4 = mm).**

OPEN CHANNEL FLOW METERS—PRIMARY DEVICES. Primary devices commonly used to measure open channel flow include Parshall flumes, Palmer–Bowlus flumes, and open flow nozzles. Although these devices all exhibit a loss of head, this is usually less a concern in pumping station design than in pure gravity systems.

Parshall flumes consist of a converging section, a throat section, a diverging section, and a stilling or measurement well. There is a decrease in level as water passes through the flume. Flow is nominally proportional to the three-halves power of the level change. *Instrument Engineers' Handbook—Process Measurement* (1982) states that this device is not extremely accurate but is adequate for wastewater applications. A commonly assumed accuracy is ±5% of rate (Briggs, 1993).

Parshall flumes are dependable, require minimum maintenance, and tend to be self cleaning because high velocities flush out solids. Available throat widths range from 25 mm to 3 m (1 in. to 10 ft). Flow range for a given flume can be as high as 35:1, but in practice the secondary's range limits the flow range to much less.

Typically, Parshall flumes are designed for free-flow (no backwater) conditions in the diverging section. As flow increases, backwater develops in the diverging section, which eventually impacts accurate measurement. In some instances, such as retrofits constrained by existing hydraulics, backwater conditions may be unavoidable at certain flows. If such flows are of interest, level measurement in the diverging and the converging section must be measured. This two-point measurement adds complexity to calculating flow; a few small vendors specialize in this type of measurement.

The Parshall flume should not be installed near turbulent, surging or unbalanced flow, or near a poorly distributed velocity profile. It should be located without bends immediately upstream.

A drawback of Parshall flumes is their expense, primarily in the concrete work needed for the flume itself and the additional concrete work needed upstream to ensure a smooth transition.

Figure 7.10 shows a Parshall flume, with free-flow conditions indicated in Table 7.1. An ultrasonic sensor measuring flow depths at H_a and H_b converts the reading to a flow value. Figure 7.11 shows an admittance-type secondary used in open channel flow measurement. The sensor strip is mounted to the side of the open channel by means of a mounting track.

The Palmer–Bowlus flume is suitable for measuring flow in round bottom open channels and is trapezoidal. Units are generally available with an integral approach section. Its advantages and disadvantages are similar to the Parshall flume. Available sizes range from 100 to 750 mm (5 to 30 in.). It can measure slightly lower flows than the Parshall flume, but the Parshall is capable of measuring higher flows.

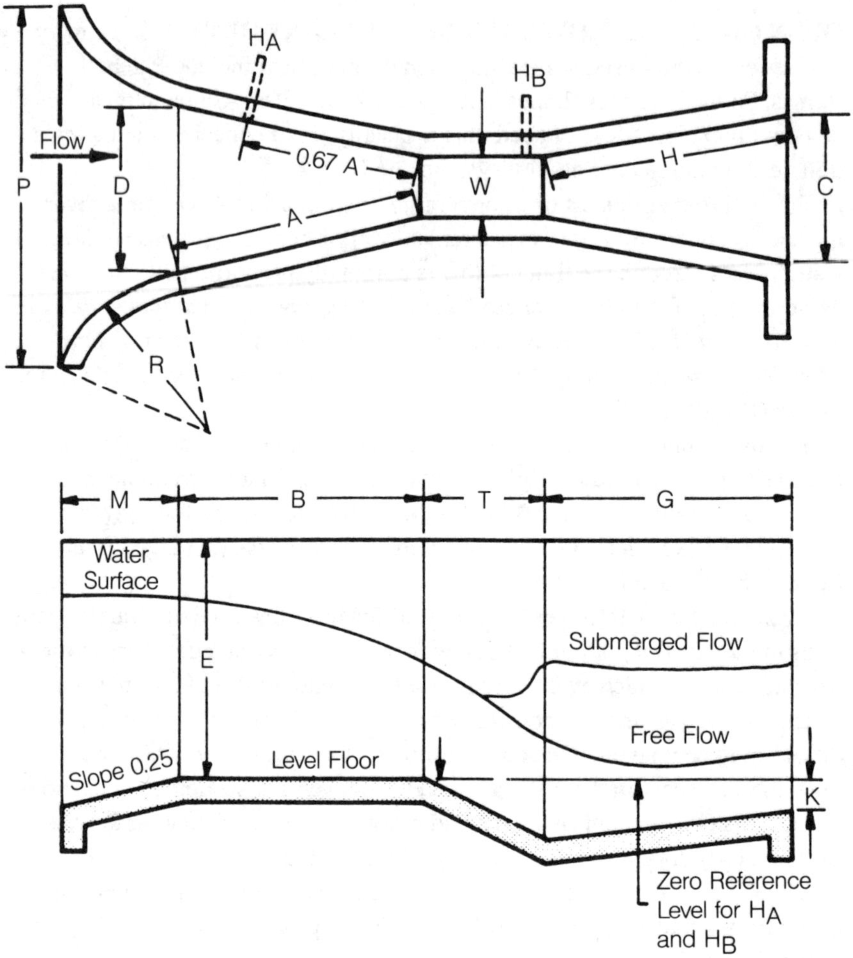

Figure 7.10 Parshall flume (WEF, 1992).

Table 7.1 Free-flow conditions for a Parshall flume (WEF, 1992).[a]

Throat width, cm[a]	Maximum submergence $H_a/H_b \times$ 100 for free flow condition, %
≤ 7.6	50
15.2	60
22.9	60
$> 22.9 < 30.5$	70
≥ 30.5	67

[a] cm × 2.54 = in.

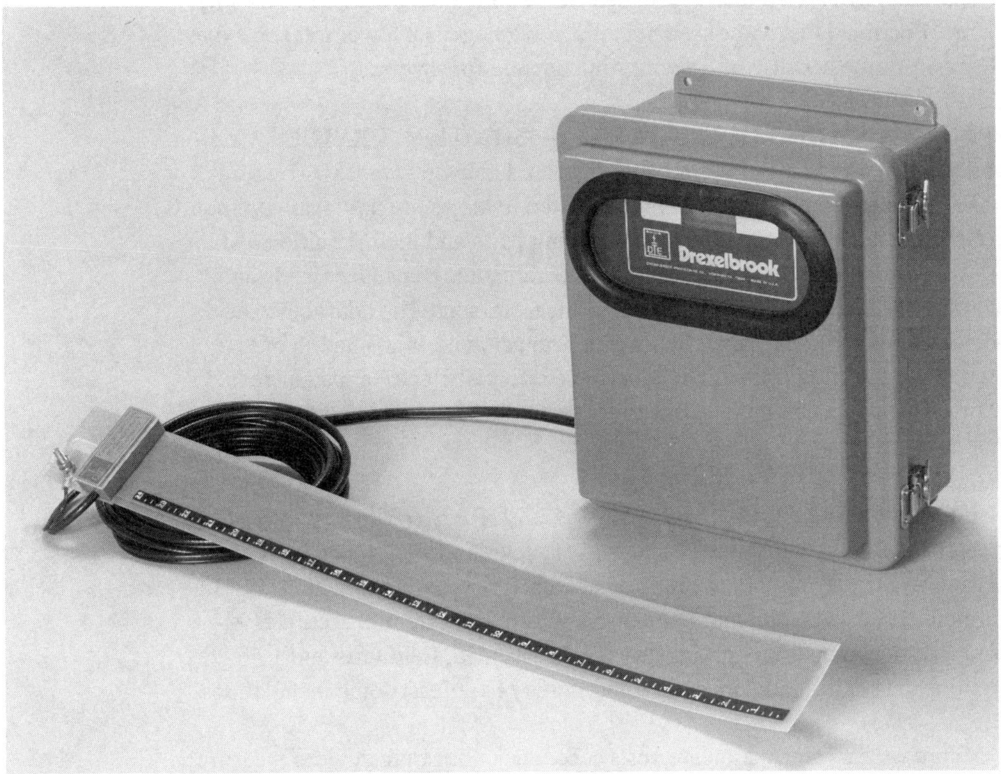

Figure 7.11 Admittance-type secondary—open channel flow measurement. Courtesy of Drexelbrook Engineering Company.

Weirs are the most accurate of these devices, but are seldom used in wastewater or stormwater systems because of their tendency to allow solids and debris to settle out at the weir base.

Flow nozzles include the Kennison nozzle and at least one proprietary device. These devices measure flow in open channels or partially filled pipes. The devices are metal with a flanged circular inlet and there is free discharge. The proprietary device has an optional transition section which allows connection to a downstream pipe.

Nozzles are at least available from 150 to 910 mm (5 to 35 in.) and with associated flow spans from 125 to 875 L/s (200 gpm to 20 mgd). Stated accuracies are at least as good as for Parshall flumes.

These devices have an intermediate section whose geometry causes a change in water surface level, which is the basis of flow measurement. Secondaries for these devices include ultrasonics, admittance probes, float systems, and even pressure-sensing systems. Required straight upstream diameters and other installation requirements should be reviewed with the manufacturer.

These devices are similar to Parshall flumes, including their self-cleaning feature. For some installations, a potential advantage is a lower total cost over a Parshall flume because of a savings in concrete formwork.

OPEN CHANNEL FLOW METERS—SECONDARY DEVICES. Common types of electronics that measure level and convert it to flow are ultrasonics and admittance devices. Float-actuated linkages are now less common, probably because of their reliance on moving parts and their requiring a stilling well, which tends to collect solids. The ultrasonic transducer is virtually identical to that used for ultrasonic level measurement. The admittance device transmits radio frequency to measure capacitance as a function of level. The device has circuitry, which reduces the effect of electrical paths created by process buildup on the probe. Proper grounding is a must and a grounding plate is usually provided.

OPEN CHANNEL FLOW METERS—ACCURACY. The accuracy of open channel flow meters is open to some speculation. For weirs, with proper precautions, the relation between flow and level is accurate to within 2% (Instrument, 1982). Parshall and Palmer–Bowlus flumes are less accurate. Although an accuracy of approximately 5% is assumed, field tests and unpublished studies indicate greater inaccuracy in at least some installations (Briggs, 1993).

Manufacturer information suggests secondary ultrasonic devices measure level with accuracy greater than 1%. Equivalent information for secondary admittance devices is more difficult to find, but is probably comparable. Field observations and common sense suggest open channel flow measurement is often more inaccurate than commonly assumed.

At most, head height is only 0.4 m (1.25 ft) and is often much less. An error in liquid measurement of, for example, 13 mm (0.5 in.) impacts accuracy substantially. Potential sources for error are numerous. Foam (even slight) in wastewater and debris collected along walls and edges can change the surface profile. Wind can produce ripples. Other possible error sources include excessive approach velocities, deviations in installing the flume liner, and inlet flow whose profile has still not transitioned from pipe to open channel flow.

Flow inaccuracy is the sum of the inaccuracies of the devices that comprise the flow-measuring system. For open channel flow meters and differential producers, one must account for the inaccuracies of the primary and secondary elements. The secondary might be miscalibrated and can drift. Admittance devices appear prone to some drift in wastewater applications, possibly from coatings buildup. Periodic maintenance is necessary.

Pumping stations do have the benefit of a wet well. If removing the pumping station from service is permitted, the wet well can be used to verify meter accuracy through drawup calibration tests.

ANCILLARY FLOW METERS. Other flow meters are used to support ancillary functions of the wastewater or stormwater pumping station. Rotameters and devices that visually indicate only flow presence are used on small flows such as for seal water. Rotameters are also used to measure air flow in bubbler systems. Propeller meters are often used for applications that require larger amounts of clean water such as washdown water.

PRESSURE MEASUREMENT

Compared to water distribution pumping stations, pressure measurement is of less concern in wastewater and stormwater pumping stations.

Pressure gauges are usually furnished on the discharge and suction of each pump. An isolation valve typically is installed with the gauge or seal to allow easy removal. Placing gauges on both pump suction and discharge assist in determining the actual field pump curve. Any pressure gauge on the pump suction should be provided with a combination vacuum or pressure range. Pressure switches are less prevalent, and are used for such applications as monitoring seal water pressure and pump discharge pressure (to confirm the pump is operating).

Pressure gauges typically are selected so that indication is between 30 and 70% during normal operation. Ideally, the pressure gauge should indicate less than 100% of full scale at shutoff head. Often both conditions are not possible and a compromise must be reached. Most pressure gauges can be overpressured approximately 30% without needing to recalibrate the gauge.

Occasionally only one gauge per pumping station is used. In such cases, process connections (either plugged or furnished with an isolation valve) should be installed on both pump suction and discharge to facilitate adding pressure gauges if they are found needed.

Pressure gauges tend to fail prematurely from mechanical shock from hydraulic transients and fast-closing valves. Accordingly, gauges are furnished with pulsation dampeners, are liquid filled, or are remotely mounted from the process pipe.

Pressure gauges typically are fitted with quarter turn isolation valves. To extend gauge life, the valves should be closed unless a reading is being taken.

For protection from debris and the process fluid itself, gauges are often furnished with diaphragm seals filled with an inert fluid. Typically the seals themselves are sufficient, but as a supplement, water purges on the process side of the seal can also be used. A suitable diaphragm material is an appropriate stainless steel alloy. A 13-mm (0.5-in.) National Pipe Thread process connection on the seal generally is satisfactory. If excessive debris or grease is anticipated, 20 or even 25 mm (0.75 or 1 in.) process connections can be used.

Two devices (gauge plus switch) mounted on one seal is common practice. Mounting three (or more) devices on one seal should be reviewed with the manufacturer. The diaphragm seal fill fluid may be unable to provide enough displacement volume. A slow-responding or inaccurate device may result, and the physical mounting of three devices may prove troublesome.

An alternative to diaphragm seals are "donut shaped" annular seals mounted off the main process line. Typical size (of the "donut's hole") would be 25 mm (1 in.) Female National Pipe Thread, and these seals can be flushed clean by opening an isolation valve. Mounting annular seals off line is still not common, thus experience is limited. Traditionally, annular seals have been mounted inline in the main process line. Because wastewater and stormwater are relatively low in solids, mounting annular seals inline probably is overdesign. For larger pipe diameters, cost becomes a factor because annular seal size is the same as the process pipe.

Although not routinely used in wastewater and stormwater pumping stations, pressure switches occasionally have some application. Normally, process pressure conditions are relatively low [less than 345 kPa (50 psig)]. Unfortunately, many standard commercially available switches in this range have appreciable deadbands, which can lead to nonreproducible behavior and control strategy malfunction (Gemin, 1993). In these cases, either a nonstandard expensive pressure switch must be used, or the control strategy may need to be abandoned altogether.

Figure 7.12 shows an installed pressure gauge and diaphragm seal assembly, with the diaphragm seal directly below the gauge. Figure 7.13 shows noninstalled annular seals. The smaller seal would be mounted off line the process pipe. The larger annular seal would be mounted inline, but would not typically be used in wastewater and stormwater applications.

ALARMS AND INDICATION

ALARMS. In small pumping stations, only abnormal high and low levels might be alarmed. Many existing small pumping stations do not alarm abnormal low level. However, this practice is not recommended because it provides no alert that a pump is running dry.

For larger sophisticated pumping stations, alarms can become numerous, and their use should be evaluated carefully. Many designers provide for monitoring multiple alarms locally at the pumping station, with only a single station trouble alarm sent to a central facility.

The numerous conditions that might be alarmed at larger pumping stations are shown in Table 7.2. Personnel safety pushbutton alarm stations have been installed no more than 50 mm (24 in.) above the floor to allow an injured operator to request help.

Figure 7.12 Pressure gauge and diaphragm seal assembly.

Pump failure is primarily detected by high liquid level in the wet well. In more sophisticated pumping systems, secondary methods to detect pump failure also are used.

There are several common secondary methods of detecting pump failure. In each, the control system monitors that the pump has been proved to be on within a preset (but adjustable) time. If no proof is received, a pump failure is assumed. Although not a definitive proof, a relatively troublefree means of sensing pump operation are the "M" contacts from the pump motor starter. More definitive but still a secondary means of sensing a running pump include check valve limit switches, flow monitoring of the pump discharge, and pump discharge pressure. Check valve limit switches generally are reliable, as long as the check valve is so designed. Field installing limit switches on check valves not designed to be furnished with such is problematic. Suitable, reasonable-cost flow switches for "dirty water" applications are hard to find. If used, pressure switches should be located on the discharge between the

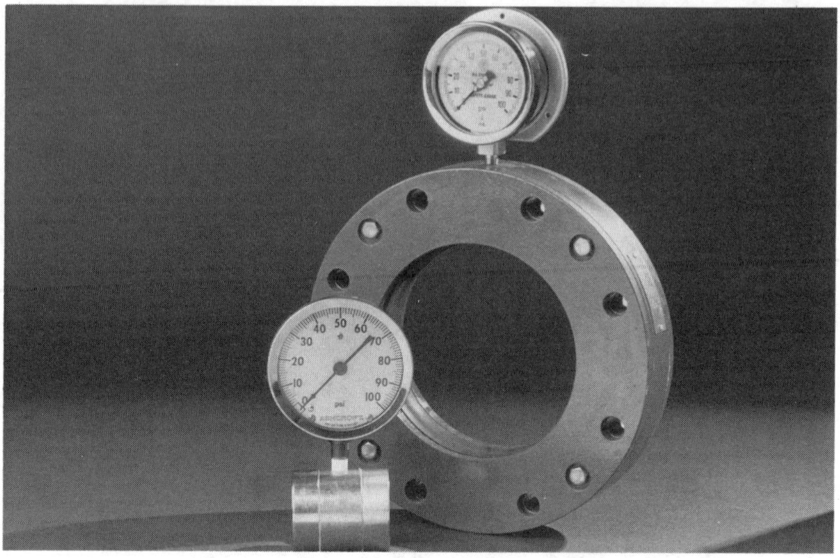

Figure 7.13 Pressure gauge and annular seal assembly. Courtesy of Red Valve Co. Inc.

Table 7.2 Selected alarm conditions at wastewater and stormwater pumping stations.

Condition

- (Each) pump fail to run[a]
- (Each) pump low seal water pressure
- (Each) check valve fail to open
- High wet well level
- Low wet well level
- High discharge flow
- Low discharge flow
- (Each) pump high vibration
- (Each) pump high bearing temperature
- Moisture present (within submersible pump motor)
- Combustible gas present
- Oxygen deficiency
- Variable frequency drive diagnostic
- Remote terminal unit failure
- Surge valve open

Condition

- Low ventilation airflow
- Building security
- Building drain sump high level
- High motor temperature
- Power failure
- Emergency generator in operation
- Intrusion alarms on perimeter fencing
- (Each) check valve fail to close
- High building temperature
- Low fuel
- Phase failure
- Motor/breaker failure
- Surge fail to close
- Building flooded

[a] See text for ways of sensing pump fail.

pump and check valve. Sometimes the correct pressure setpoint is difficult to determine. Contributing factors include wide changes in suction pressure, low dynamic friction pipe losses, and multiple-pump operation that changes normal operating pressure.

Often, hand switches that control pumps are three-position switches (ON/OFF/AUTO). These fail conditions typically are interlocked only with the AUTO switch position, otherwise nuisance alarms might occur if the switch is operated in the ON position during testing or maintenance.

A possible exception to this rule is if the fail condition can cause injury or excessive mechanical damage. For these critical fail conditions, some designers choose to provide an alarm and an interlock to stop the pump even in the ON mode. An example is stopping a progressive cavity pump on high pressure.

Atmospheric monitoring devices typically are installed in pumping station working areas. The requirements of NFPA 820 (National Fire Protection Association, 1992) should be followed, but the party that usually needs to be satisfied is the local fire marshall, whose requirements may differ. National Fire Protection Association (NFPA) 820 requires that combustible gas detectors be placed in rooms or enclosed spaces of both wastewater and stormwater pumping stations, and that they be of the fail safe type. Also, NFPA 820 requires that loss of ventilation flow be monitored. In general, NFPA 820 requires that these alarms be both visual and audible. Other devices often installed include combustible gas and oxygen deficiency detectors.

EQUIPMENT STATUS INDICATION. Status of most major process equipment is indicated at the pumping station. Typical status indications include those for pumps (constant speed), variable-speed drives, check valves, screening and degritting equipment (if present), ventilation fans, station power and emergency generator. Occasionally, the pumps commanded to run in the automatic sequence (lead or lag) are also indicated.

Except for smaller pumps, elapsed run time of each pump since installation is usually indicated at the pumping station or at the central facility. The run time and number of starts since midnight (or some other time) might also be indicated.

Above a certain motor size, for example, 37 kW (50 hp), motor amperage draw, voltage to each of the three legs, and motor speed (in revolutions per minute) typically are indicated, particularly with variable-speed drives.

PROCESS VARIABLE INDICATION. Depending on pumping station and overall system complexity, the number of variables measured can be none at all to several. Some guidelines apply. Flow, wet well level, and totalized flow normally are indicated at the pumping station. Flow is more often recorded than level.

The use of recorders at the pumping station sometimes is problematic. Recorders need pen and chart paper replacement, and are prone to attack by corrosive gas associated with wastewater pumping stations. Accordingly, recorders should be installed in a protected atmosphere so as to be reliable and long lasting. If located outside, recorders are subject to environmental attack such as by sunlight and heat. To reduce environmental effects, recorders should be furnished with a rain and sun shield and should be installed facing north.

For supervisory control and data acquisition (SCADA) systems, the central facility typically has a report printer and desktop monitor. The report printer sometimes is omitted, which can be a problem if a hard copy of reports is wanted.

OPERATOR INTERFACES

Operator interfaces at the pumping station site include a control panel and hand switches near the pump. At remote central facilities for SCADA telemetry systems, the operator interface typically is a desktop workstation with keyboard.

The face of the control panel at the pumping station typically includes hand switches, status indicating lights, alarm lights or an annunciator, and process variable indicators and recorders. Operator interface units typically are provided on the panel face for devices such as single loop digital controllers and programmable controllers. For simple stations that use programmable controllers, an operator interface unit may not be provided. Instead, a personal computer occasionally is used to access the programmable controller to examine registers, modify strategies, and adjust parameters.

CONTROL PANELS. Control panels should be constructed with the appropriate NEMA rating. Table 7.3 describes National Electrical Manufacturers Association (NEMA) classifications for control panels typically used in wastewater and stormwater pumping stations. The listed panels are rated for use in nonhazardous locations; panels rated for hazardous use seldom are required in wastewater and stormwater pumping stations.

Because NEMA 12 panels do not resist corrosion well, the more conservative and appropriate choice for indoor process areas in wastewater pumping stations is NEMA 4X (watertight and corrosion resistant). Along with NEMA 1, NEMA 12 might be appropriate in control and in electrical rooms. Where high corrosion is expected or has been experienced, NEMA 4X panels have been used. For outdoor applications, the conservative choice is NEMA 4X panels. In areas which are not humid and with small amounts of corrosive sewer gases, NEMA 3, NEMA 3R, and NEMA 3S have proved adequate.

Table 7.3 NEMA classifications of control panels commonly used in wastewater and stormwater pumping stations.[a]

Enclosure	Description
NEMA 1	Indoor use—some protection against contact with the enclosed equipment
NEMA 3	Outdoor use—protection against windblown dust, rain, sleet, and external ice
NEMA 3R	Outdoor use—protection against falling rain, sleet, and external ice
NEMA 3S	Outdoor use—protection against windblown dust, rain, sleet; operation of external mechanisms when ice laden
NEMA 4	Indoor or outdoor use—protection against windblown dust and rain, splashing water, and hose-directed water
NEMA 4X	Indoor or outdoor use—protection against corrosion, windblown dust and rain, splashing water, and hose-directed water
NEMA 12	Indoor use—protection against dust, falling dirt, and dropping noncorrosive liquids

[a] Adapted from ANSI/NEMA 250, Enclosures for Electrical Equipment (1985).

The corrosion test that a NEMA 4X panel must pass involves subjecting the panel to a salt spray. Thus, NEMA 4X panels should not be considered universally corrosion resistant. Common enclosure materials for NEMA 4X panels include aluminum, fiber glass, 304 stainless steel and 316 stainless steel.

NEMA 4X panels add a complication in that many analog operator interface devices such as indicators and recorders are not similarly rated. Accordingly, Plexiglas™ windows must be provided on the panel front for these devices. Corrosion-inhibiting capsules should be placed within the control panel, though a dehumidification system also may prevent corrosion.

For outdoor installations, solar shields should be used on control panels in high solar energy environments and within which are housed sensitive solid state electronics. Instead of solar shields, some practitioners have painted the control panel exterior white to reflect solar radiation. Such control panels should be placed with their face in the northerly direction to reduce heat buildup and to allow easy viewing of low-intensity light-emitting device indicators. Even in relatively warm areas, control panels located outdoors should have thermostatically controlled heaters to prevent condensation.

Control power is typically 115 V–ac. In some applications, low-voltage (24-V–ac or dc) control power is used instead. Examples of these applications include standby engines or certain variable-frequency drives (VFDs), for which low-voltage control power is the manufacturer's standard. Such low-voltage control circuits should be used with caution. Electronic devices in low-voltage control circuits appear more prone to induced voltages from nearby process equipment power—460-V, three-phase in the U.S.; 575-V, three-phase in Canada. The induced voltages sometimes render low-voltage control relays nonfunctional. "Band-aid" field fixes such as using interposing relays may work, but the interposing relays themselves require 120-V–ac power.

Figure 7.14 shows a stainless steel NEMA 4X control panel located outdoors.

Figure 7.14 NEMA 4X control panel.

ANCILLARY INSTRUMENTATION. Ancillary instrumentation includes devices such as elapsed time meters and process variable indicators, recorders, and totalizers. The enhanced capabilities of microprocessor-based circular chart recorders make them attractive. Multiple process variables can be recorded and otherwise processed. Chart ranges are custom printed by the recorder itself. Also, process-variable totalizers are provided integral to the recorder.

CONTROL SYSTEMS

Types of control systems for automatic pump operation include conventional control relays, programmable logic controllers, single loop digital controllers, proprietary controllers, and "smart" remote terminal units (RTUs).

Distributed control systems seldom are used in wastewater and stormwater pumping stations because they are suited for large, complicated processes. The application of personal computers and minicomputers typically is reserved for use at central facilities in SCADA.

The choice of control system depends on cost and control complexity and flexibility. The control systems typically are located at the pumping station and, if there is a central control facility, it generally performs "supervisory control" only with real-time control implemented by controls at the pumping station. Occasionally, central facility control is the normal mode, with reversion to local control only if the central facility control system fails.

CONVENTIONAL CONTROL RELAYS. Conventional control relays are still frequently used and are dependable and easy to obtain, even locally. Ladder logic and implementing logic changes with control relays usually can be learned without taking specialized courses.

Simple programmable controllers, acting only as relay replacers, sometimes have a lower installation cost than relays, though relays might still be appropriate in some cases. For instance, most practitioners would not want multiple pumping stations where each control system was composed of a different model (sometimes) obscure programmable controller lacking both proper documentation and programming units.

As control schemes become more complex, the amount of wiring and number of relays becomes cumbersome. Making field changes is time consuming, and programmable controllers become appropriate in these instances.

PROGRAMMABLE LOGIC CONTROLLERS. Programmable logic controllers have become increasingly prevalent because they are field programmable and implementing control logic is flexible. Properly installed programmable logic controllers have demonstrated high reliability.

Programmable controllers were originally developed as relay replacers in the automotive industry. Discrete sequencing logic is their forte, and programmable controllers are still typically programmed in ladder logic. In the past decade, their ability to manipulate analog variables has increased to the point where they are trusted to perform relatively simple analog control functions. For more complex analog control functions, the amount of software programming and documentation becomes cumbersome.

For pumping station applications, programmable controllers typically start and stop pumps and control pump speed as a function of wet well level, alternate the pumps in sometimes complicated schemes, totalize flows, and monitor for fail conditions. Programmable controllers benefit from power conditioning, either with a power conditioner or an uninterruptible power supply (UPS), in which the designer wants the programmable controller to function during power failure. Most programmable controllers are rated for operating at temperatures between 0 and 60°C (32 and 140°F) and relative humidities between 5 and 95% without condensation. Thus, programmable controllers can be installed in areas other than conditioned control rooms as long as precautions (such as installing control panel heaters or solar shields) are taken. Programmable controllers have been known to halt their operation when exposed to too much sun or when located within a building during hot summer days.

Programmable controller output modules are classified into two basic types: a voltage output that can drive devices or an isolated contact output, which is a newer development. Proponents of either type can be found.

A suggested practice in using voltage output modules is to wire them directly to controlled devices, such as relays, which reside in the same control panel as the programmable controller. If the signal is to leave the panel, interposing relays should be used, which cuts down outside electrical interferences.

Isolated contact outputs can be directly wired into control circuits outside the panel and are available for driving up to NEMA 5 starters. The advantage of the isolated contact outputs is that they eliminate relays. However, the module itself is more costly. Also, because this module is larger, there is little if any savings in panel size. Besides a maximum current rating when driving a load, isolated contact outputs also have a minimum required current, presumably to extend contact life. This minimum current rating should be accounted for in the design.

SINGLE LOOP DIGITAL CONTROLLERS AND OTHER DEVICES. Single loop digital controllers (SLDCs) are also field programmable and typically are suited for applications where less control logic is required. Though more suited for analog control functions, SLDCs also can provide discrete sequencing. Some SLDCs use a menu of logic blocks to implement the algorithm. Certain practitioners prefer this type of "language" over ladder logic.

Several proprietary control systems are also available. In one, a microprocessor-based controller uses preprogrammed software routines to control pumps for situations normally encountered in wastewater and stormwater pumping stations. Routines include ramping adjustable-speed pumps as a function of wet well level and various alternation schemes.

Supervisory control and data acquisition systems communicate with pumping stations by means of RTUs located at the pumping stations themselves. In years past, the only function of the RTU would be as part of the communication link. Local logic was done by other devices such as control relays or PLCs. "Smart" RTUs are increasing in popularity. Because local control algorithms reside in these devices, the design is simplified.

CONTROL ALGORITHMS

Control algorithms must achieve the process objective and also must not expose the pump to harmful operating regimes. Pumping stations can be composed of constant-speed pumps, variable-speed pumps, or a combination of the two.

The simplest process objective is to pump the wet well contents to a receiving point without overflowing the wet well. Constant-speed pumps and their algorithms are appropriate here. Also, some wastewater plants prefer there always be some minimum flow to the plant. This is sometimes accomplished with a small, constant-speed "jockey" pump. As diurnal flows increase, the normal pumping scheme is invoked.

Often, the receiving point prefers accepting a varying flow not subject to step increases. An example is a large pumping station that discharges to a small wastewater plant.

Applicable algorithms require at least some variable-speed pumping capacity. Such variable-speed capacity is provided with all variable-speed pumps, a combination of variable- and constant-speed pumps, two-speed motors, or algorithms that make use of different size pumps.

Control algorithms must also guard against potential harmful effects on pumps. For instance, as level rises and a large capacity, high head pump starts, a smaller capacity, low head pump might be interlocked to prevent it from pumping against its shutoff head and incurring damage.

There are numerous control schemes, equipment and logic for pumping applications. A thorough analysis of the pumping station is required to confirm the most reliable and cost-effective system to use. The following are some guidelines.

ALL CONSTANT-SPEED PUMPS. In the most general case of an all-constant-speed pumping station, each pump is provided with unique start and stop level setpoints. A simple variation typically used on small pumping stations is to start multiple pumps individually on successive higher wet well levels, but to stop them at the same level. As pumping station size increases relative to the receiving wastewater plant, this variation places unnecessary flow and load surges on the plant and should be avoided.

Another variation is to use time delay logic coupled with the same level setpoint to control multiple pumps. For instance, a lead pump will start on a rising high level, and the lag pump will start if high level is sustained for a preset time, such as 1 minute. Flow conditions and wet well volume should be carefully examined to ensure the normal operating range of the wet well is not exceeded.

Equal-size pumps with two-speed motors can be controlled to produce an output comparable to adjustable-speed pumps.

Occasionally, two different size constant-speed pumps are controlled to produce similar results. Consider the case of three pumps, each with single-speed motors. One has a nominal capacity of 1Q, and two have nominal capacities of 2Q. As level rises, pumps can be started and stopped to produce pumping station nominal outputs of 1Q, 2Q, 3Q, 4Q, and 5Q. Wet well volume should be analyzed to ensure pump-starts-per-hour criteria are not exceeded.

ALL VARIABLE-SPEED PUMPS. The simplest case of variable-speed pumping stations is one active pump. Two setpoints (high and low) will dictate start and stop of the pump. Pump speed typically will ramp up from minimum to maximum speed as level rises from low to high. An equivalent control scheme uses proportional mode feedback control with preset gain and initialization factors. The minimum speed setting must be set to keep pump operation within the pump operating curve, thus is determined by the system head-capacity curve. For centrifugal pumps, this minimum speed setting is typically approximately 70 to 80%.

Proportional integral (two mode) control of a level setpoint has been attempted but, in general, is not recommended. Although successful in certain applications, this scheme has potential problems. If tuned too tightly and if flow varies rapidly, pump speed will oscillate wildly in attempting to maintain the setpoint level.

For multiple, identical, variable-speed pumps, each pump starts and stops via unique setpoints. A less common alternative starts and stops multiple pumps via a total of only two setpoints used in conjunction with time delay logic. A common practice is to operate all pumps at the same speed in a load share mode which maintains the same capacity for each pump at the system head. As the next pump starts, the speed of all running pumps is reduced from maximum to some lower speed (not necessarily minimum pump speed). For instance, in a two-pump case, the lower speed is selected such that the flow discharge of the two pumps is slightly less than the output of one pump at maximum speed. This provides an overlap, preventing possible excessive cycling between the one pump on at maximum speed versus two pumps on at a lower speed.

Although the load share mode is common, algorithms sometimes are developed so that two pumps run at different speeds, that is, staggered arrangement. A typical reason for such algorithms is to permit the use of a common variable-speed drive for two pumps (Gottliebson, 1977).

The speed at which each variable-speed pump runs typically is determined by the ramping curves that are part of the overall control algorithm. Each ramping curve typically provides pump speed as a function of wet well level. For multiple pumps operating in a load-share mode, there may be a single or multiple ramping curves.

If there are multiple ramping curves, the ramping curve followed at any instant typically depends on wet well level or the number of pumps on or both. Multiple ramping curves can each occupy separate vertical segments of the wet well or the vertical segments of the curves can overlap. The latter practice makes good use of wet well volume and especially is useful if the designer wants to reduce wet well depth.

Figure 7.15 presents the control sequence for the three active pump, variable-speed pumping system shown in Figure 4.25. The pumps run in a load-share mode. Only one ramping curve is used, though often multiple ramping curves are used instead.

COMBINATION OF CONSTANT- AND VARIABLE-SPEED PUMPS. Particular care is required in controlling stations having a combination of variable- and constant-speed pumps. It is important to keep the variable-speed pump within its operating envelope. A careful analysis is required, in part because the total dynamic head (TDH) of the variable-speed pump changes with pump speed, which is related to wet well level.

Consider a pumping scheme with constant- and variable-speed pumps. At certain wet well levels, the variable-speed pump may be at or near shutoff head because of the constant-speed pump. Possible damage to the lag pump from radial thrust forces may result. A possible solution narrows the speed range of the variable-speed pump so its range of TDH is compatible with that of the constant-speed pump. However, narrowing the speed range reduces pumping station flow range, which may be unacceptable from a process standpoint.

Figures 7.16 and 7.17 present example control algorithms of a pumping station comprised of both constant- and variable-speed pumps.

The Figure 7.16 algorithm has no discontinuities or "gaps" in the pumping station flow output as level rises in the wet well. The pumping station consists of one lead 63-L/s (1 000-gpm) variable-speed pump and one lag 32-L/s (500-gpm) constant-speed pump. The variable speed operating range is 3.33 to 1, or 20 to 63 L/s (300 to 1 000 gpm).

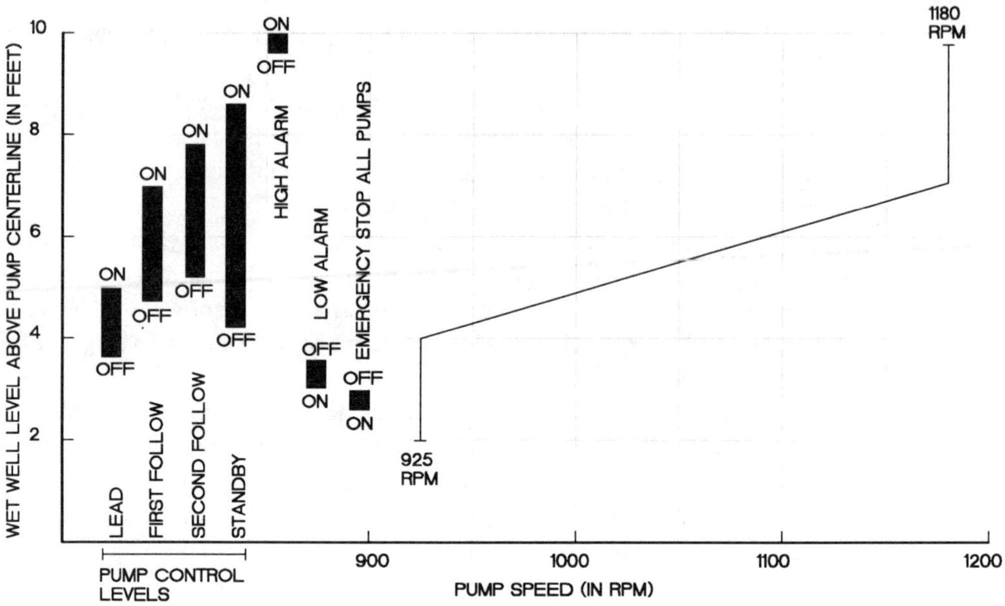

Figure 7.15 Control algorithm for pumping system of Figure 4.25 (ft × 0.304 8 = m).

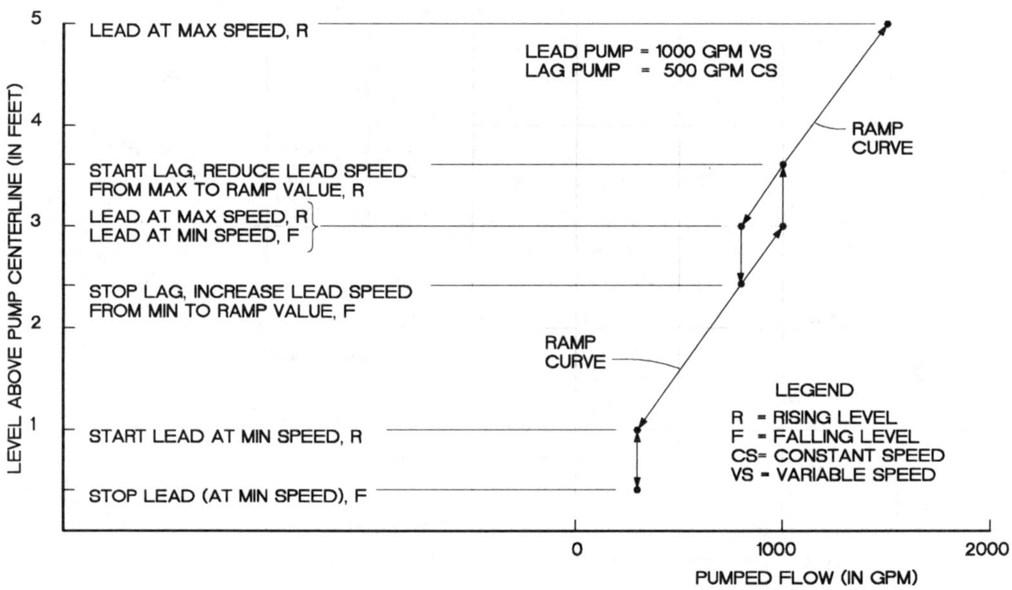

Figure 7.16 No gaps in control algorithm—combination of constant- and variable-speed pumps (ft × 0.304 8 = m and gpm × 6.308 × 10^{-5} = m^3/s).

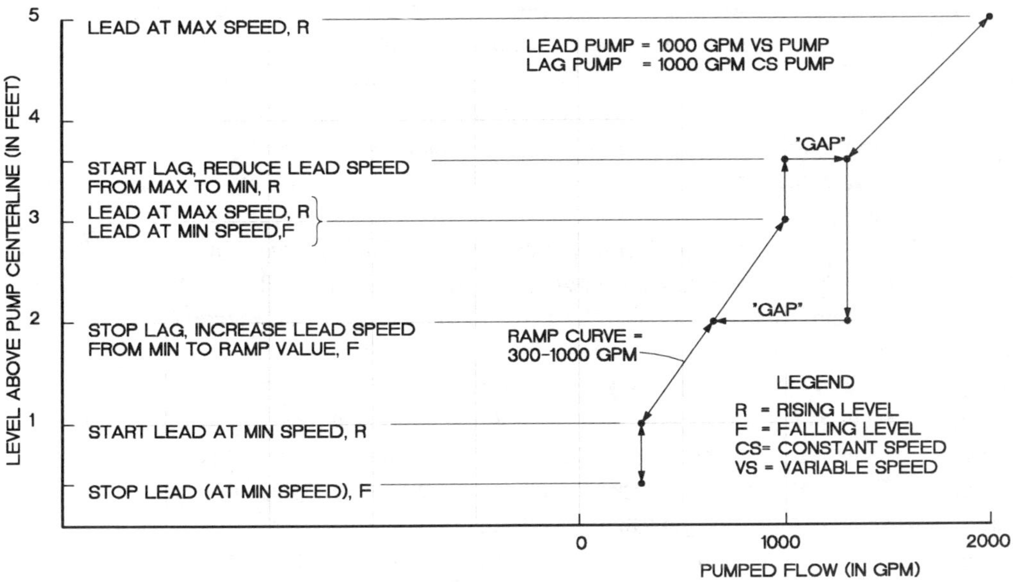

Figure 7.17 Gaps in control algorithm—combination of constant- and variable-speed pumps (ft × 0.304 8 = m and gpm × 6.308 × 10^{-5} = m^3/s).

As level rises in the wet well, the lead variable-speed pump eventually reaches 63 L/s (1 000 gpm). As level continues to rise, the 32-L/s (500-gpm) lag pump starts. Once the lag pump starts, the lead variable-speed pump reduces or "ramps down" its speed so its output is also reduced to 32 L/s. Total pumping station output remains at 63 L/s, assuming a slow rise. There is no flow discontinuity or "gap." Control algorithms such as this generally have few problems associated with them.

The Figure 7.17 algorithm exhibits a flow discontinuity or "gap" as level rises in the wet well. The pumping station consists of one lead 63-L/s (1 000-gpm) variable-speed pump and one lag 63-L/s (1 000-gpm) constant-speed pump. The variable-speed pump range is 3.33 to 1, or 20 to 63 L/s (300 to 1 000 gpm).

As level rises in the wet well, the lead variable-speed pump eventually reaches 63 L/s (1 000 gpm). As level continues to rise, the 63-L/s lag pump starts. The lead pump is "ramped down." However, at its minimum speed, its output is not zero but 20 L/s (300 gpm). Thus, the pumping station output jumps from 63 to 83 L/s (1 000 to 1 300 gpm) for an incremental change in level, and the pumping station has a 20-L/s "gap."

If inflow to the pumping station remains at 63 L/s (1 000 gpm), the pumping station will rapidly pump down and the lag constant-speed pump will soon stop. This scenario promotes possible excessive cycling for the installed motor horsepower.

Gaps usually occur when there is insufficient variable-speed pumping capacity during a constant-speed pumping transition. Although gaps in such algorithms can cause major cycling problems, some practitioners successfully use them. As long as "gaps" can be tolerated, equal-sized, variable- and constant-speed pumps can be used, which is an advantage. Such advantages might include interchangeable parts, ease of wet well layout, and ease in pump selection.

A 1 135-L/s (18 000-gpm) peak flow pumping station that has "gaps" in its algorithm has been in successful operation for more than 8 years (Bergs, 1993). The pumping station consists of four equal-sized pumps, with two constant-speed pumps and two variable-speed pumps. Of the two variable-speed pumps, one is active and the other is a standby. Gaps do not appear to pose a problem for several reasons. The wet well is large enough to keep pump cycling time within recommended limits. Flow to the pumping station is either very low or very high. Thus, flows that would accentuate adverse "gap" behavior are infrequent, and the gaps themselves are not particularly large.

In summary, algorithms with "gaps" can produce both excessive cycling and produce large flow swings to the receiving entity. If "gap" algorithms are to be used, the pumping station should be designed with caution and thoroughness to reduce possible cycling problems.

ALTERNATION. Algorithms are also developed to alternate pumps. Pump alternation exercises the pumps, resulting in evenly distributed pump wear. However, there is a school of thought that equal pump wear promotes simultaneous pump failure and should be avoided.

Various hardware exists to alternate and select lead and multiple lag pumps in pumping stations. These include electromechanical alternators, electromechanical stepping switches, microprocessor-based alternators, lead and lag selector switches for manual alternation, and manual alternation by plug and jack systems, typically reserved for use for sequences with many pumps. Programmable controllers can also be programmed to cause alternation.

Alternation schemes range from simple to complex. An example of a simple alternation scheme is to alternate the lead pump in a duplex pump station. Complex automatic alternation schemes include first on and first off, and last on and first off, with initiation by time clock. The first on and first off scheme turns off the pump with the longest run time during the previous cycle, and starts the pump with the longest rest time. Another complex scheme provides uneven run time on pumps. Even run time concerns some practitioners in that pumps may all fail at the same time.

For pumping stations routinely visited by operators, the most cost-effective scheme is to operate the pumps in a fixed sequence periodically changed by the operator with a selector switch or jacks.

Common ways to automatically initiate alternation include reaching low level when all pumps are off, at periodic real-time intervals, or when elapsed time of a lead pump exceeds a preset value.

Sometimes, automatic alternation is used in the retrofit of an undersized existing wet well because alternation effectively increases the time between starts of an individual pump. For instance, consider a duplex constant-speed pumping station with a required firm pumping capacity of 1Q. Each pump has a capacity of 1Q. Starter and motor sizes are such that they require 30 minutes between pump starts.

At critical inflow (Q/2) to the existing wet well, the one firm pump might have a minimum time between starts of only 15 minutes. However, by alternating pumps, the time between starts of any individual pump is an acceptable 30 minutes. Because acceptable starts per hour require that two pumps must be available for operation, pump capacity is not a truly firm 1Q. To rigorously make this capacity firm, another 1Q pump must be installed. The designer must decide if the total installed cost of this arrangement is less than the cost of enlarging the wet well.

ABNORMAL OPERATING CONDITIONS. Abnormal operating conditions often pose problems, which must not be overlooked in designing the control system. The control system should provide an orderly start-up after a power outage.

To reduce hydraulic transients and simultaneous motor inrush currents, pump starts often are staggered by the use of appropriately set time delay relays. By contrast for some systems, hydraulic transients unexpectedly are greatly reduced when pumps start up simultaneously because pressures that slam check valves are balanced out. However, the electrical system would experience additive motor inrush currents.

Protective control circuits should be analyzed to ensure they enable the process equipment to automatically restart after an outage.

For control systems with programmable controllers, its processor and input/output peripherals should be capable of resuming operation after an outage or "brownout." Typically, this is feasible with some minor reconfiguration.

As previously stated, programmable controllers benefit from power conditioning. Some designers choose to use an uninterruptible power supply (UPS) to provide conditioned power. Others view the UPS as unnecessary because the programmable controller is only of use with main power available to the pumps. A UPS will reduce problems associated with "brownouts." However, if a UPS is used, the control algorithm must be carefully reviewed to ensure proper process equipment operation after a power outage. For example, the UPS-backed programmable controller may sense a false fail of process equipment during a power outage. Once power is resumed, the "failed" equipment might erroneously be interlocked from restarting.

Often for its own protection, a pump is shut down on an abnormal low level and provisions must be made to restart the pump, though manual reset generally is not recommended. An appropriate method is to restart the pump automatically as level rises to some higher level. This introduces deadband, which reduces excessive pump cycling.

Excessive pump cycling can prevent proper operation of a pumping station. In one instance during low inflow, a remote pump was forced to shutdown because of high motor temperature developed by excessive cycling on an abnormal low level. Once inflow resumed, the station overflowed because the pump could not restart until manually reset.

VARIABLE-SPEED DRIVES

Variable-speed pump drive systems also are used for reasons other than smoothing flows to a receiving point. Variable-speed pump drive systems also reduce the number of motor starts, which increases motor life and reduces power demand charges. The number of motor starts also reduces transients to the pumping and piping system, which extends mechanical equipment life.

Variable-speed pump drive systems may also offer potential energy savings, but this typically is more of a factor in higher pressure water distribution systems. Potential energy savings are more achievable in systems with more frictional losses relative to static lift.

Because required wet well volume is less, variable-speed pump drive systems and their algorithms are also used in either retrofitting an undersized wet well or constructing a wet well from scratch to save construction costs. Variable-speed drives are covered in more detail in Chapter 6.

MISCELLANEOUS

INSTALLATION. Proper installation of instrumentation and other control components sometimes is overlooked. Fortunately, several firms are starting to specialize in supplying installation accessories, which should facilitate good installation practices. These commercially available accessories include such items as insulated enclosures and heating systems for field instruments to winterize them, modular design instrument supports, sunshades, and preinsulated (and traced) tubing bundles.

Recommendations of suppliers should be followed, but sometimes this is not completely possible. For instance, the preference for installing most flow meters is vertical with upward flow, but mechanical layouts often preclude this. Required straight upstream diameters for some flow meters can be substantial. At least one manufacturer of Venturi-type flow meters accommodates less than ideal conditions by providing the amount of accuracy loss if the amount of straight upstream pipe is less than recommended.

Structural support for instruments and control panels should resist corrosion. Stainless steel (316 preferred) or aluminum members are recommended, as are stainless steel fasteners.

Because of their weight, control panel support should be reviewed closely with wind force taken into account. The temptation to handrail mount instruments and control panels generally should be resisted. Though sometimes overlooked, OSHA regulations require the top handrail be unobstructed, increasing installation complexity. Especially if rain hoods are provided, the wind force can damage the handrail. Hanging components on handrails also reduces free clearance.

Because of its corrosion resistance, stainless steel tubing for applications such as between differential pressure primary elements and transmitters is probably worth the extra expense over copper. In cold areas, tubing should be heat traced or provisions made for water purges. Solar shades should be provided for instruments in direct sunlight with temperature sensitive components (most solid-state devices).

In areas with wide temperature swings of moist air, condensation in control panels and conduits can occur. Panels should be equipped with panel drains and breathers. Condensation in conduit containing the cable between primary element and its transmitter can cause problems, though sealing conduits with caulk might help. Figure 7.18 details how the condensate can be allowed to drain while still maintaining the watertight integrity of the conduit.

SEAL WATER. Several types of centrifugal pumps require seal water; there are several seal water control systems to consider. For a pumping station in water-rich areas with relatively inexpensive pumps, continuous-seal water flow even with the pumps off might be considered. However, in the opinion of some practitioners, this practice should be avoided in that water should always be conserved.

The seal water solenoid valve often is interlocked with the pump control circuit to supply water only when the pump is on. Fail open (deenergized to open) solenoids are preferred because seal water will be supplied to the solenoid even if power to the solenoid is lost. For expensive pumps, monitoring seal water by pressure switch and subsequent pump shutdown may also be warranted.

SURGE SUPPRESSION. Where lightning strikes are possible, surge suppression on system components should be considered. In lightning-prone areas, surge suppression is strongly recommended. Even in areas not lightning prone, some experienced practitioners recommend surge suppression.

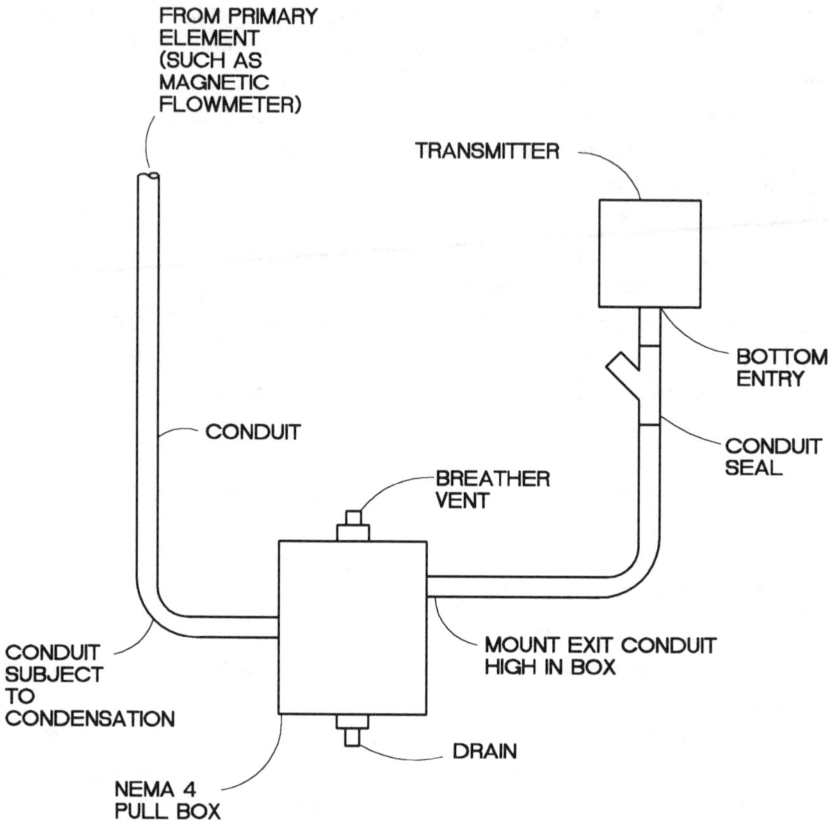

Figure 7.18 Conduit condensate drain detail.

The designer must exercise judgment in determining which equipment needs surge suppression. In lightning-prone areas, surge suppression should be provided for ac power to control panels and transmitters, communication lines (telephone lines for telemetry systems and antenna cable for radio and microwave transmission systems), and control and analog signal wiring to devices located in another building or outdoors.

The designer must also decide on the type of surge suppressor. In lightning-prone areas, ac power and analog signal lines should be furnished with fast-response surge suppressors with low clamping voltages such as the double stage hybrid type. These devices are available in NEMA 4X enclosures, which facilitate protecting transmitters located outdoors. In other areas, the designer again can use the double stage hybrid type, but the lower cost metal oxide varistor or gas discharge suppressors may be appropriate instead. Proper grounding is essential for these devices to operate effectively.

CONTRACT DOCUMENT COORDINATION. The instrumentation and control (I&C) system is integrated in the sense that several subcontractors or vendors have components which either are part of or are controlled by the I&C system. Any of these weak links can break the I&C system "chain."

Coordination problems during construction are common, but these coordination problems can be reduced by using judgment in assigning I&C components to a particular section of the specifications.

In large, complicated pumping stations, project execution often is assigned to an I&C systems integrator. Such integrators provide a variety of services, including field instrumentation purchase and field calibration, fabrication of control panels or equivalent, programming and configuration services, and testing. They typically have trained field personnel who often provide on-the-spot, trouble-shooting services. They also produce thorough and well-organized documentation of the I&C system, which is of great value.

Pump vendors sometimes provide control systems as part of their standard package. These control systems are either done in house or done by contract by another firm. Control requirements of the project should be checked against the logic provided as part of the standard package. If much different, an I&C systems integrator should be considered. Sample documentation of previous package control systems should be reviewed during design to check for quality. If quality appears to be a problem, an I&C systems integrator again should be considered.

Smaller projects sometimes do not justify the use of an I&C systems integrator, but rather the control system is provided by the pump vendor or even by the electrical subcontractor. Some field instrumentation still is required for these projects. If instrumentation is minor such as just a few float switches, they can be furnished either by the pump vendor or the electrical subcontractor. More care should be given for more complicated components such as Venturi tubes. Whatever section such an item is specified in, the specification should include requirements for field services, including instrument calibration and field testing.

The I&C system is closely related to the electrical subsystem, and the specifications should be coordinated so that adequate documentation is required by subcontractors to facilitate construction. Panel elementary drawings that include terminal numbers must be provided. Interconnecting drawings must be provided that show wiring (including intermediate terminal junction boxes) between control panels and motor control centers.

TRAINING AND MAINTENANCE. Many control systems of pumping stations do not function well after a time because of poor maintenance. Field instruments should be periodically calibrated, inspected, and repaired. Hardware and software control systems problems should be diagnosed and corrected.

Maintenance can be done in house or a maintenance contract can be procured with a systems integrator. If done in house, instrument technicians, electrical technicians, or trained individuals with an interest in control systems can perform the required maintenance and diagnosis. Some but not all of these skills can be learned on the job. Formal training courses are beneficial.

Training courses include topics such as instrument calibration and programming, and the design of large projects often include formal training of some operations personnel. Ideally, the owner should make I&C training an ongoing program.

TELEMETRY SYSTEMS

Monitoring of remote pumping stations often is done at a central facility, and sometimes central facility control of the pumping stations also is provided. Telemetry systems link remote pumping stations and the central facility. Regardless of the telemetry system used, the design of the central facility control system should provide for any expected future expansion.

USER NEEDS. Central facility monitoring keeps personnel aware of operating conditions at remote pumping stations. Such knowledge should reduce the number and quantity of spills by providing a quicker emergency response. Noncritical abnormal operating conditions can be ascertained and corrected before a critical condition develops.

Though the above reasons are not directly related to cost savings, for some owners they offer enough justification for a telemetry system.

Some telemetry systems are justified on a savings of operations costs, such as energy savings and the labor savings that result from reduced trips to inspect and acquire data from remote pumping stations. Telemetry systems are particularly justified where overflow reporting is required within a set time limit by a regulatory permit. A single fine can approach the cost of a telemetry system (Peterschmidt, 1993).

As a minimum, several key abnormal conditions should be monitored from each pumping station. These include abnormal high levels; a lost communication link between the central facility and the remote pump station; and a loss of power to the remote pumping station (Weirens, 1992). Another condition almost always worth monitoring is a general trouble alarm from the remote pumping station. Besides monitoring, the central facility can also be configured to control these remote pumping stations.

In stormwater and wastewater system, central facility control is required when systemwide management is practiced, for example, flow storage or routing to different wastewater plants for peak shaving or equalization. Sometimes central facility control is used to increase reliability of remote pumping stations. Thus, if local control at the pumping station malfunctions, the central facility assumes control. For such cases, the control systems must be carefully examined to ensure central facility override control is independent of local control.

SYSTEM TYPES. Telemetry system types include automatic telephone dialers and tone and digital telemetry. Supervisory control and data acquisition systems can be built using many telemetry system types. However, digital telemetry systems lend themselves to complex SCADA functions such as archiving data. Current SCADA systems of at least moderate complexity invariably are digitally based.

Automatic telephone dialer systems are often used in smaller systems. The automatic dialer uses a dial up line, an advantage because of its low cost. On an alarm condition such as a high wet well level, the automatic dialer calls a preprogrammed number with a voice message. Calls are placed to such locales as the central facility, an answering service, a pager assigned to an "oncall" employee, or directly to an "oncall" employee. Most dialers have battery backup so they can function during a power outage, but systems with multiple dialers can experience problems with common systemwide alarms, for example, a power failure.

Automatic dialer capabilities are growing to include messages that give process variable information, pump run time, station totalized flow, and even some SCADA functions. These include starting and stopping pumps and printing out at the central facility all abnormal conditions for a pumping station in alarm.

Some dialers have had problems placing long distance calls. Lightning surges have also caused dialers to issue false alarms (Ouzts, 1992). The dialer itself should be protected with a surge suppressor.

Tone telemetry conveys information over a communication link by means of audio tones. The two basic types of tone telemetry include amplitude modulated (AM) and frequency shift keyed (FSK). A given AM tone telemetry device uses only a single carrier frequency and thus conveys only two-state information (on–off). Thus, a nonmultiplexed AM transmitter or receiver would have only 1 Form C contact. It is difficult to detect a loss of carrier with AM tone telemetry.

Though more versatile and secure from noise than AM tone telemetry, FSK tone telemetry is more expensive. A given FSK tone telemetry device uses a single carrier center frequency and frequency drifted mark and space frequencies to convey information. Thus, a nonmultiplexed FSK channel can have up to 3 Form C contacts, one for each of the carrier, mark, and space frequencies. One of the contacts can be used to detect communications failure.

The communication link for tone telemetry is typically a leased telephone line. Both discrete and analog signals can be communicated. As a minimum, status and alarm discrete signals are sent from the pumping station to the central facility. In some systems, the central facility also sends supervisory control signals to pumping stations. Some systems also send analog signals to the central facility to provide pumping station flow and wet well level information.

Multiple analog and discrete signals from a single pumping station can be transmitted over a single telephone line using multiplexers that interface with the tone telemetry equipment. Multiplexing is done either by time division multiplexing or multiple tone channels (frequency division multiplexing). The time division multiplexers use digital techniques that add to the security of the data.

Because of their flexibility and low cost, digital telemetry systems now are more common than tone telemetry systems in new installations. Accordingly, digital telemetry systems are integral to the design of SCADA systems.

SUPERVISORY CONTROL AND DATA ACQUISITION SYSTEMS. Supervisory control and data acquisition systems are used anywhere from medium-sized, straight forward pumping systems to large complicated ones in which there is a desire for systemwide management. At the central facility, the SCADA system provides interactive displays for status monitoring and control, alarm processing, trending of analog variables such as flow, report generation, and historical data collection.

Most SCADA systems today are microprocessor-based, using minicomputer and personal computer systems. A typical, simple SCADA system includes a single master computer with a monitoring screen and a report and alarm printer. It should be cautioned that a single computer SCADA system often contains possible single points of failure.

Sophisticated SCADA systems might include multiple computers linked together on a local area network (LAN) or a "hot standby" configuration. Multiple computers are primarily provided for redundancy. There is some flexibility in assigning tasks to each computer, though even SCADA systems with multiple computers on a LAN can contain a single point of failure (Peterschmidt, 1993).

In some systems, a satellite facility at a different location augments the central master facility. A typical communication link between the master and the satellite would be a leased telephone line. Figure 7.19 is a block diagram for a SCADA system.

The SCADA system communicates with the RTU of each individual pumping station by either of two basic methods, polling or contention. Some vendors use a combination of both. In polling, the central host periodically polls the pumping station RTUs to acquire data. In contention, RTUs only communicate with the central host if there is a change in state. A common problem with contention occurs if many RTUs detect a simultaneous failure, such as systemwide power failure. This can overload communications channels, and manual intervention is required for the system to recover (Peterschmidt, 1993).

In the design of SCADA systems, the appropriate polling time should be examined. Often, the initial desired polling time pushes the capabilities of equipment commercially available at a reasonable cost (Ouzts, 1992).

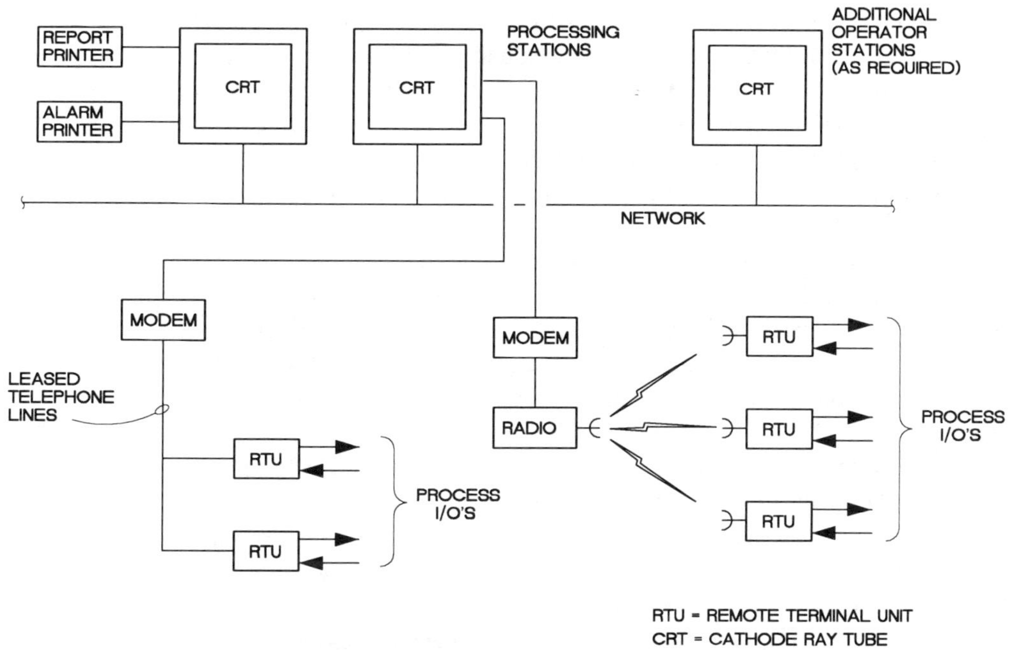

Figure 7.19 Supervisory control and data acquisition system block diagram.

A slower polling time may be more appropriate and realistic. Alternatively, a contention protocol or multiple common channels should be considered.

Even with SCADA systems, primary control typically resides at the pumping station itself, external to the SCADA system. The SCADA system could be designed to override the primary control mode, either by operator manual intervention or by altering the automatic strategy. For large sophisticated systems, logic would be distributed between the pumping station and the central facility. At the central facility, systemwide management decisions would be made. In some SCADA systems, primary control resides at the central facility. If a communication failure is sensed, control reverts to local automatic or manual control.

COMMUNICATION LINKS. A variety of communication links are available that convey information between the central facility and remote pumping stations. Several possible links include radio channels, leased telephone circuits, leased metallic telephone pairs, dial up telephone networks, cellular telephone systems, cable television systems (broadband network), microwave radio links, and satellite channels. Some links such as leased metallic telephone pairs (which use dc continuity circuits) are less available as phone companies phase them out of service. Others are appropriate only under special circumstances. Two of the most commonly used are telephone (leased circuits and dial up networks) and radio channels.

Dial up telephone networks have monthly charges that typically are much less than leased telephone circuits. Thus, where limited amounts of data are transferred and where brief delays can be tolerated in conveying information, dial up telephone networks might be preferred over leased telephone circuits.

With deregulation of the telephone system in the 1980s, leased telephone lines became more expensive on average, making the radio communication link more prevalent than previously.

Typically, radio communication link systems have a higher initial construction cost than telephone links, but lower operating costs. Typically, radio systems can reach the break even point in 3 to 5 years. The telephone leasing cost is also subject to escalation. Because the owner has less control over the telephone link, there may be difficulties in obtaining timely maintenance and trouble shooting.

If considered for use, the radio communication link must be both technically feasible and available. Suitable radio frequencies are in short supply in many populous urban and suburban areas, where licenses needed to operate the radio communication link may not be available. The first priority is to determine if licensable frequencies are available.

Topography must be considered in determining technical feasibility. Flat areas where there is a line of sight between the central facility and remote pumping stations promote radio communication. Radio "repeating stations" are used as part of the radio communication link in more rugged areas but add cost and complexity to the link.

In evaluating a radio communication link, a study phase is conducted to determine the radio system requirements and its economic feasibility. Available radio frequencies are determined and a path analysis or propagation study is performed.

Several firms offer frequency-coordination services to determine available frequencies in the proposed areas of operation. Frequency bands commonly available are midband, highband, very high frequency (VHF), ultra high frequency (UHF), and 900 MHz. These bands each have their own technical advantages and disadvantages. For instance, VHF bands are subject to "skipping," where under certain atmospheric conditions, the radio waves "skip" off atmospheric layers and can travel up to hundreds of miles. The bands are also subject to different regulatory constraints.

Based on these and other factors, a feasible available frequency can be selected for license. During the design phase, firms complete the necessary paperwork for the user to file for a Federal Communications Commission (FCC) license. Upon license award, the user has one calendar year to get the system operational or the license is revoked. Extensions can be requested, though are not routinely awarded in some popular frequency bands.

During the study phase, an initial propagation analysis should be performed to determine the parameters necessary for reliable operation. This analysis can be based on available line of site between nodes. Personal computer programs are available to perform path analysis based on latitude and longitude coordinates of proposed sites in conjunction with topographic databases.

Propagation calculations and field tests can be used to verify or reject marginal or questionable paths. Propagation is a term that describes how radio wave intensity varies in space. Reliable link communication is only possible when the strength of the received signal is easily distinguished from noise. Factors affecting propagation include the selected frequency; distance between sites; topography; obstructions, including foliage and buildings; antenna types and heights; cable lengths and types; and transmitter power and receiver sensitivity.

Fade margin is a measure of signal strength in excess of that required to just distinguish the correct signal. The expected fade margin is determined as part of the propagation study. Each link should have a positive fade margin, for example a value of 20 dB.

Typical problems encountered with radio communication links include momentary interferences from sunspots and thunderstorms. Long-term interferences include "skip" and "bleed over" from splinter frequencies.

Figure 7.20 shows a remote pumping station furnished with a radio antenna.

Figure 7.20 Remote pumping station with radio antenna.

REMOTE TERMINAL UNITS. The use of "smart" RTUs is becoming more widespread than the conventional RTU, whose only purpose is to interface signals with the communication link. Because local control algorithms reside in a "smart" RTU, the design is simplified.

Programmable controllers furnished with appropriate modems can also be used as RTUs. This practice is common and cost effective if the programmable controller is also providing the local "stand alone" control.

Lower cost alternatives to SCADA systems are pumping stations furnished with a type of "smart" RTU that only periodically (for example, daily) communicates to the central facility by telephone dial up (Morey, 1992). Information conveyed includes pump starts, pump elapsed run times, cycle times, and combined flow. These systems can also call up with unscheduled reports on an alarm condition.

Regardless of the type RTU used, it must be interfaced with the local control system. The actual operator interface is usually on the local control panel. Simple systems may not have a local control panel. Instead the operator interface would be on the motor control center itself.

RELIABILITY

Reliability is paramount for stormwater and wastewater pumping stations used in critical applications. Accordingly, reliability should be incorporated to the design of their control systems. For instance, a major wastewater pumping station that, if it failed, would surcharge sewers must be designed for reliable operation. Similarly, reliable operation is required for those stormwater pumping stations that, in the event of failure, would flood roadway underpasses, tunnels, basements, and low-lying towns protected by levees.

LEVEL SENSING ELEMENTS. Alarm sensing devices should be provided independent of normal operating sensing devices. For instance, floats can be provided to alarm abnormal levels if the bubbler system used in normal operation fails.

For level sensing systems used in normal operation, dual sensing devices and transmitters could be provided, such as two bubbler systems or ultrasonic meters provided to measure wet well level. Bubbler systems could also be provided with dual compressors.

CONTROL HARDWARE AND ALGORITHMS. The normal mode control algorithm can be supplemented by backup algorithms such as starting one or more pumps on an abnormal high level.

In pumping stations with both constant- and variable-speed pumps, often only the constant-speed pumps are configured to start on the abnormal high level. To plan for this, the designer provides enough wet well volume to prevent excessive pump starts and often includes the surcharged influent pipe as usable volume.

The heart of the control system could be provided with redundancy. For example, the main programmable logic controller might be backed up with a "hot" programmable logic controller. Also, a single loop digital controller or a group of relays may back up critical parts of the control algorithm resident in the programmable logic controller. Transfer of control from the programmable controller could be automatic by means of a watchdog timer, which senses if the programmable controller fails.

For SCADA systems, manual override operation sometimes is performed from the central facility. Also, for some SCADA systems, dual communication links such as radio and leased telephone lines may be provided. Some "smart" RTUs now have backup dial up telephone capabilities. If the "smart" RTU senses no poll, it dials up the central facility by exception reporting.

CONTROL POWER. The control panel could be designed to accept control power from two different sources. The system can be designed so that backup algorithms are powered directly from the motor control center control circuit of each individual piece of process equipment. Thus, these backup algorithms still would function if the control panel were to lose power.

Critical pumping stations often are provided with backup power sources such as a diesel generator. By design, power to all pumps may not be provided while under a backup power source. If control panel power happens to be tied in to the bus that feeds the nonpowered pumps, the control algorithm would not function.

PUMPS AND DRIVES. The type of pump used should be examined. Because screw pumps can pump dry without incurring harm, their backup control algorithms are inherently simpler than those for centrifugal pumps. For instance, the programmable controller can be configured to fail such that the pump command-to-run contacts close on power or controller fail, causing the screw pumps to run continuously.

Some variable-frequency pump drives are available with full-voltage starters so the pump drive can be run at constant speed. The switch over to the full-voltage starter can be accomplished either manually by means of a selector switch or automatically upon failure of the variable-frequency drive.

TESTING

For all but the most simple pumping stations, a formal test program starting at the factory is suggested for several reasons. The I&C system is comprised of customized, sometimes complex, logic. Misinterpretation and incorrect implementation of the design intent is possible.

Parts of the design intent may also be flawed. Many aspects of the control algorithm deal with high pumping and abnormal operating conditions, which might not be encountered by the pumping station for years unless a specific test is conducted.

Finally, for many projects, the contractor is rushing to meet contractual deadlines. Often there is only a narrow window to commission the I&C system because it depends on completion of other trades (such as mechanical and electrical).

A formal testing program lends an air of rationality under these circumstances and reduces the tendency to assume the I&C system is correctly functioning.

At least four types of tests are appropriate throughout construction, including unwitnessed factory tests, witnessed factory tests, field readiness tests, and witnessed field tests. For complicated pumping stations, all four types are suggested. Unwitnessed factory tests rely on the good faith of the control systems supplier to develop and conduct appropriate tests. In an unwitnessed test, the control panel or equivalent is tested for proper wiring and testing algorithms.

In a witnessed factory test, the owner or the owner's agent (usually the design engineer) witnesses control system testing, which is conducted by the control systems supplier. Normally, the control panels or equivalent are inspected, though the bulk of the testing covers the control algorithms. Test procedures are developed in advance of the test itself and are often detailed. Testing typically involves the simulation of process conditions and deficiencies are noted and corrected.

Field readiness tests do not test logic functions. Instead, these tests verify that individual instrumentation and control components are installed and working, with wires properly terminated and set points set correctly. These tests typically are unwitnessed but must be documented to serve as a basis to proceed with the witnessed field tests.

Witnessed field testing is important because it demonstrates that the I&C system accomplishes all functions required by the design. Accordingly, passing these tests typically is linked to contractual milestones such as substantial completion and final acceptance. Testing is similar to the witnessed factory tests, except that real process variables are used as much as possible. For instance, tests may include wet well pumpdowns of water or wastewater to demonstrate proper function of the lead and lag pumps. Any deficiencies are noted and retested (either witnessed or unwitnessed).

Witnessed field testing often is conducted as other interrelated systems are completed. Thus, problems with these interrelated systems will be identified as testing of the I&C system proceeds.

To reduce testing bottlenecks, it is best to have certain trades either at the test or readily available. Two common trades of benefit would be electrical and mechanical. If too many deficiencies are identified, testing progress becomes so slow that the test will need to be postponed until the interrelated systems are made operable.

REFERENCES

American National Standards Institute/National Electrical Manufacturers Association (1985) Enclosures for Electrical Equipment. ANSI/NEMA 250-1985, Washington, D.C.

Bergs, A., RV Anderson Associates, Willowdale, Ontario, Canada (1993) Personal Communication.

Briggs, B., Primary Flow Signal, Pawtucket, R.I. (1993) Personal Communication.

Cooper, R.W., ADS Equipment, Portland, Oreg. (1992) Personal Communication.

deGlanville, B., CH2M Hill, Corvallis, Oreg. (1992) Personal Communication.

DiNapoli, L., BIF, North Wales, Pa. (1993) Personal Communication.

Gemin, J., Proctor and Redfern, Kitchener, Ontario, Canada (1993) Personal Communication.

Gottliebson, M. (1977) Explore the Use of Variable Speed Water Booster Pumps/2. *Water Wastes Eng.,* **14,** 10, 58.

Guagno, J., EMA Services, Philadelphia, Pa. (1992) Personal Communication.

Gunby, F.M., Jr., Metcalf and Eddy, Woburn, Mass. (1993) Personal Communication.

Instrument Engineers' Handbook—Process Measurement (1982). B.G. Liptak and K. Venczel (Eds.), Chilton Book Company, Radnor, Pa.

Instrument Society of America (1992) Instrumentation Symbols and Identification. S5.1 (formerly ANSI Y32.20), Research Triangle Park, N.C.

Morey, E., Atlanta, Ga. (1992) Personal Communication.

National Fire Protection Association (1992) Recommended Practice for Fire Protection in Wastewater Treatment Plants. NFPA 820, Quincy, Mass.

Ouzts, J., Turbitrol Company, Atlanta, Ga. (1992) Personal Communication.

Peterschmidt, W., CH2M Hill, Corvallis, Oreg. (1993) Personal Communication.

Robinson, C. (1986) Obstructionless Flowmeters: Smooth Sailing for Some, Rough Passage for Others. *InTech,* **33,** 12, 33.

Water Environment Federation (1992) *Design of Municipal Wastewater Treatment Plants.* Manual of Practice No. 8, Alexandria, Va.; Manual of Engineering Practice No. 76, Am. Soc. Civ. Eng., New York, N.Y.

Weirens, D.J., Consolidated Electric, St. Paul, Minn. (1992) Personal Communication.

Whitenack, W., Town of Morristown, Morristown, N.J. (1992) Personal Communication.

Suggested Readings

Dulas, J.H., and Tisdale, J.C. (1992) Improve Flow Control with Electronic Drives. *Chem. Eng. Prog.*, **88,** 2, 40.

IEEE Recommended Practices and Requirements for Harmonic Control in Electric Power Systems (1991). IEEE Stand. 519, New York, N.Y.

Instrument Society of America (1986) *Standards and Practices for Instrumentation.* 8th Ed., Research Triangle Park, N.C.

Jones, C.T., and Bryan, L.A. (1983) Programmable Controllers—Concepts and Applications. International Programmable Controls, Inc., Atlanta, Ga.

Keskar, P.Y. (1988) PLC System with Emergency Generators and Variable Frequency Drives: A Hard Knocks Education. *Programmable Controls,* **7,** 4, 36.

Occupational Safety and Health Association (1983) OSHA Safety and Health Standards. 29 CFR 1910, OSHA 2206, 1910.23(e).

Sanks, R.L., *et al.* (Eds.) (1989) *Pumping Station Design*. Butterworth Publishers, Stoneham, Mass.

Skrenter, R.G. (1987) *Instrumentation Handbook for Water and Wastewater Treatment Plants.* Lewis Publishers, Inc., Chelsea, Mich.

U.S. Department of the Interior (1984) *Water Measurement Manual.* Bureau of Reclamation, Denver, Colo.

Water Pollution Control Federation (1984) *Process Instrumentation and Control Systems.* Manual of Practice No. OM-6, Washington, D.C.

Chapter 8
Appurtenances

SCREENING

The screening of wastewater and stormwater typically is used to remove large floating and submerged objects that may otherwise damage pumps and other equipment, obstruct pipelines, and interfere with the normal operation of pumping and any downstream treatment facilities. Screening that removes oversized solids is distinguished from comminution or grinding, which reduces the size of solids without removing the solids from the wastewater. A basic design decision is whether to screen only to the extent necessary to protect the pumping equipment or whether to maximize screenings removal. While many factors enter into the decision, screening to maximize solids removal is usually practiced only at treatment plants and is discussed in detail in *Design of Municipal Wastewater Treatment Plants* (WEF, 1992). Generally, screening for wastewater and stormwater pumping stations is accomplished with coarse or bar screens.

Bar screens consist of vertical or inclined steel bars spaced at equal intervals across a channel through which stormwater or wastewater flows. Criteria used in their design include bar size, spacing and angle from the vertical, channel width, and approach velocity. The most commonly used coarse screens are mechanically cleaned bar screens, although manually cleaned bar screens in smaller pumping stations are not unusual. This section also includes comminutors because of their use ahead of pumping equipment in some installations instead of bar screens.

TYPES OF SCREENS. Coarse or bar screens, typically used as the first item at the point of influent flow to the pumping station, include trash racks, manually and mechanically cleaned bar screens, and basket screens.

Trash Racks. Trash racks are bar screens with large openings [40 to 150 mm (1.5 to 6 in.)] that are designed to prevent logs, timbers, stumps, and other large heavy debris from entering the pumping station. Trash racks, typically followed by bar screens with smaller openings, are principally used in stormwater and combined systems that carry high amounts of large debris. Trash racks may be mechanically or manually cleaned. Where space is limited, pumping stations sometimes have basket-type trash screens that are manually hoisted and cleaned.

Manually Cleaned Screens. Manually cleaned bar screens typically have 25- to 55-mm (1- to 2-in.) openings, with the bars set 30 to 45 degrees from the vertical to facilitate cleaning (Figure 8.1). Screens are typically constructed of heavy evenly spaced rectangular or round steel bars set nearly vertically in a flow channel. Bars are held in a parallel, equal-spaced position by a flat toe plate across their lower ends and by horizontal spacing bars welded to the screen's downstream side. These horizontal spacing bars should be designed so that they will not interfere with raking the screen, and the lower bar should be above the normal flow area of the screen.

Manually cleaned bar screens typically are custom designed by the engineer and specially fabricated from drawings and specifications. After fabrication, the screen is, if carbon steel, hot-dip galvanized for corrosion protection. Screens may also be fabricated of reinforced plastic, aluminum, or stainless steel.

Although manually cleaned screens require little or no equipment maintenance, they may present significant labor-related maintenance costs. They demand frequent raking to avoid clogging. Screens are manually raked onto a drainage area or perforated plate where they drain before removal or disposal. Provision should be made for a raking tool that fits the screen bar spacing and a convenient method of handling the screenings storage container. Removal

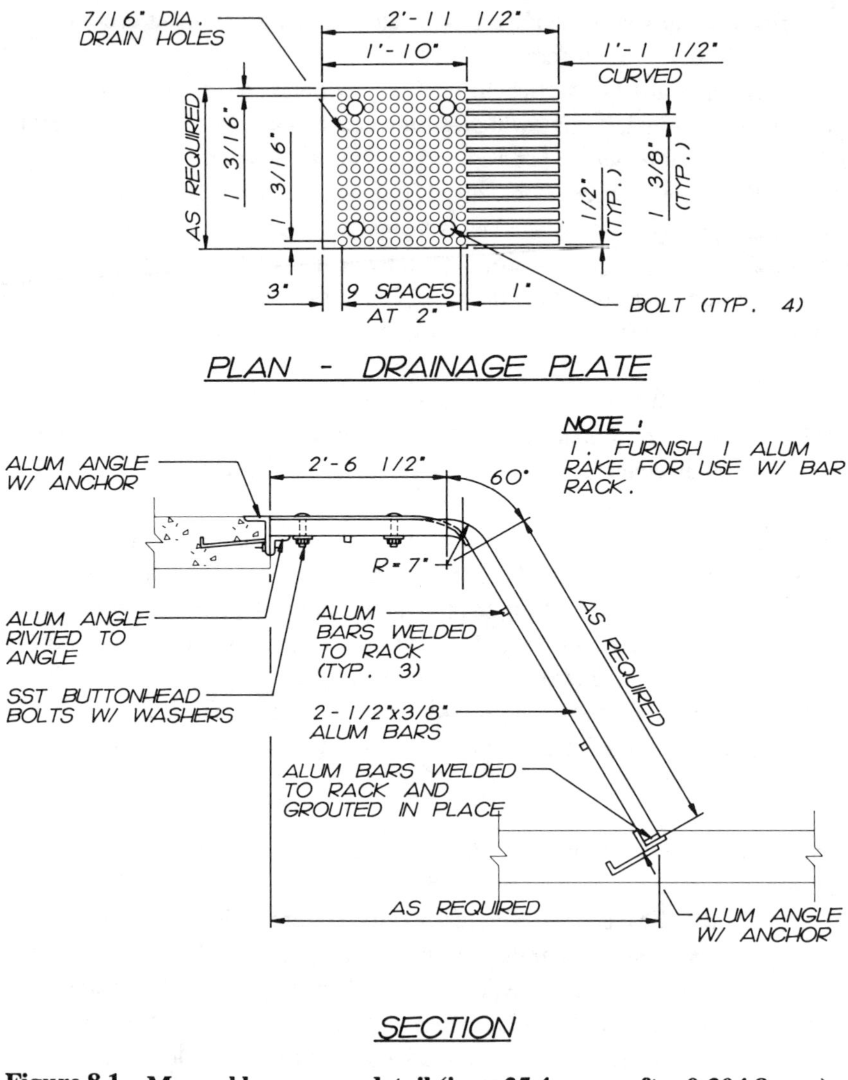

Figure 8.1 Manual bar screen detail (in. × 25.4 = mm; ft × 0.304 8 = m).

of screenings may require a hoist or dolly. Manually cleaned screens exist primarily in older, small pumping facilities and bypasses of comminutors and mechanically cleaned screens. Where entry to the wet well is required to rake the bar screen, confined space entry procedures and requirements should be followed. Because of the potentially intensive labor involved, the use of manually cleaned screens as the primary screening method should be avoided.

Basket Screens. Basket screens (Figure 8.2) are typically in the form of a box with one partially open side facing the incoming sewer or are cylindrical with an open top. If flooding of the screen is unlikely due to high wet well

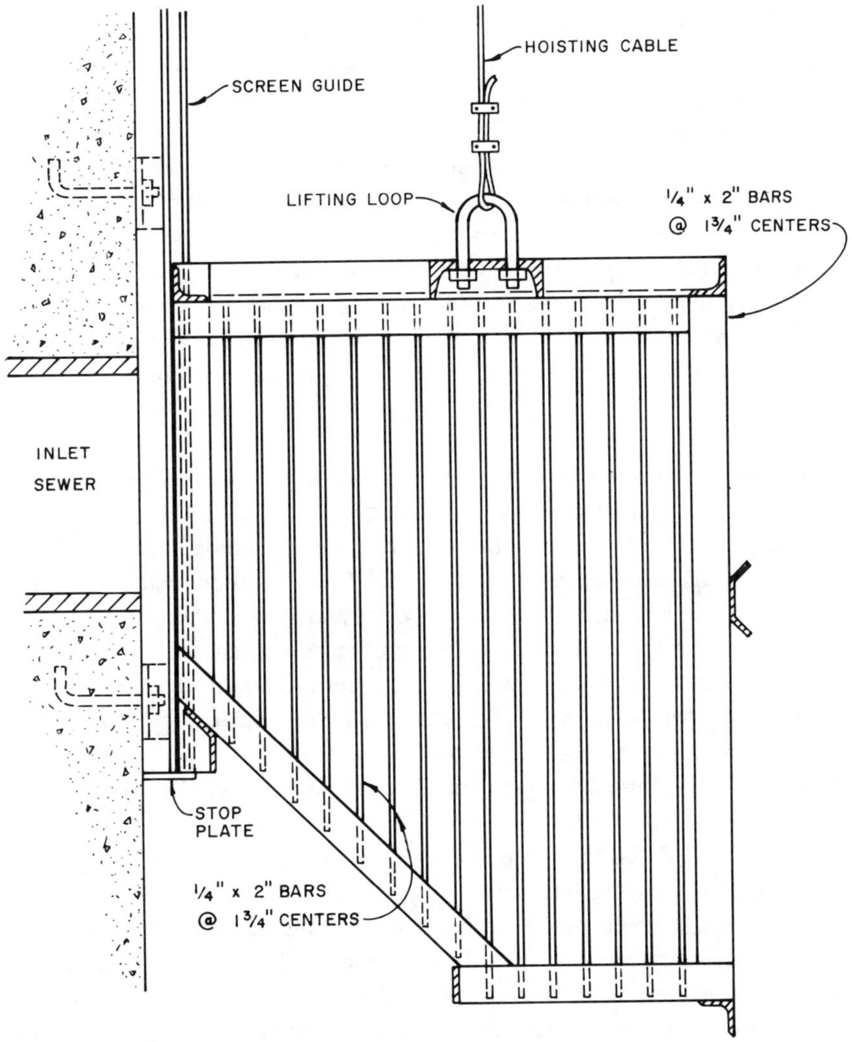

Figure 8.2 Basket screen (in. × 25.4 = mm).

water level, the top of the screen box may be left open. Parallel flat bar construction of basket screens, particularly the bottom, makes cleaning easier than does round bar construction. Grating or mesh-type construction should not be used because such screens are harder to clean. Cross members for bar spacing and structural support should interfere as little as possible with basket cleaning.

Basket screens typically are raised for emptying and lowered by means of a corrosion-resistant cable and a winch, preferably motor operated. The hoisting arrangement permits the raised basket screen to be maneuvered easily to a location for convenient removal of screenings to a storage container. Guides and supports should be designed to hold the screen in a stable position while it is being emptied. In the wet well, supports and guides should be installed so that the basket screen, when lowered, always returns to the proper position for effective screening of the discharge from the inlet sewer. Manhole steps or a ladder should be installed to permit access to the screen when it is in the operating position.

Some of the advantages of basket screens are that channel, dry well, and access facility space are not required; basket screens are well suited for submersible pump stations that may not have any dry well facilities; and screenings removal (hoisting) is a relatively simple operation.

Some of the disadvantages of basket screens are that they are subject to clogging and overflow because screen cleaning is not automated (limited capacity) and there is potential hang-up or sticking of the basket in raising or lowering.

Mechanically Cleaned Screens. Mechanically cleaned screens have openings typically ranging from 5 to 40 mm (0.25 to 1.5 in.), with bars set from 0 to 30 degrees from the vertical. Mechanical cleaning, compared with manual cleaning, tends to reduce labor cost; improve flow conditions and screenings capture; reduce nuisances; and, in stormwater and combined systems, better handle large quantities of debris and screenings. A mechanically cleaned screen is, therefore, typically specified for larger [>3 800-m^3/d (1-mgd)] pumping stations. Many types of mechanically cleaned bar screens are manufactured, including but not limited to chain- or cable-driven with front or back cleaning, reciprocating rake, rotating rake arm, catenary, and continuous.

CHAIN- OR CABLE-DRIVEN SCREENS. Mechanical bar screens with a chain or cable mechanism (Figure 8.3) to move the rake teeth through the screen openings are the oldest mechanized screening devices. These types of screens are manufactured in several configurations: front clean/front return, front clean/rear return, and back (or through) clean/rear return. The front clean/front return type most efficiently retains captured screening by minimizing carry-over. Such screens are used extensively in ordinary municipal wastewater applications. Front clean/rear return screens are used for heavy duty applications (Pankratz, 1988). A disadvantage of a rear return configuration is the possible return to the screened waste stream of debris that is not dislodged by the rake cleaning mechanism. For any of the front-cleaned configurations, bars should be trapezoidal in cross section, providing a tapered opening to prevent wedging and trapping of solids between the bars. Disadvantages of front-cleaned screens include the possibility of bottom jamming by unusual deposits of trash, particularly if no trash racks are used. Also, the chain- or cable-driven raking mechanism consisting of submerged sprockets or other mechanical devices is subject to fouling by grit and rags. Frequent inspection and maintenance of the drive mechanism is typically required and, depending on the screen design, channel dewatering may be necessary. Rake

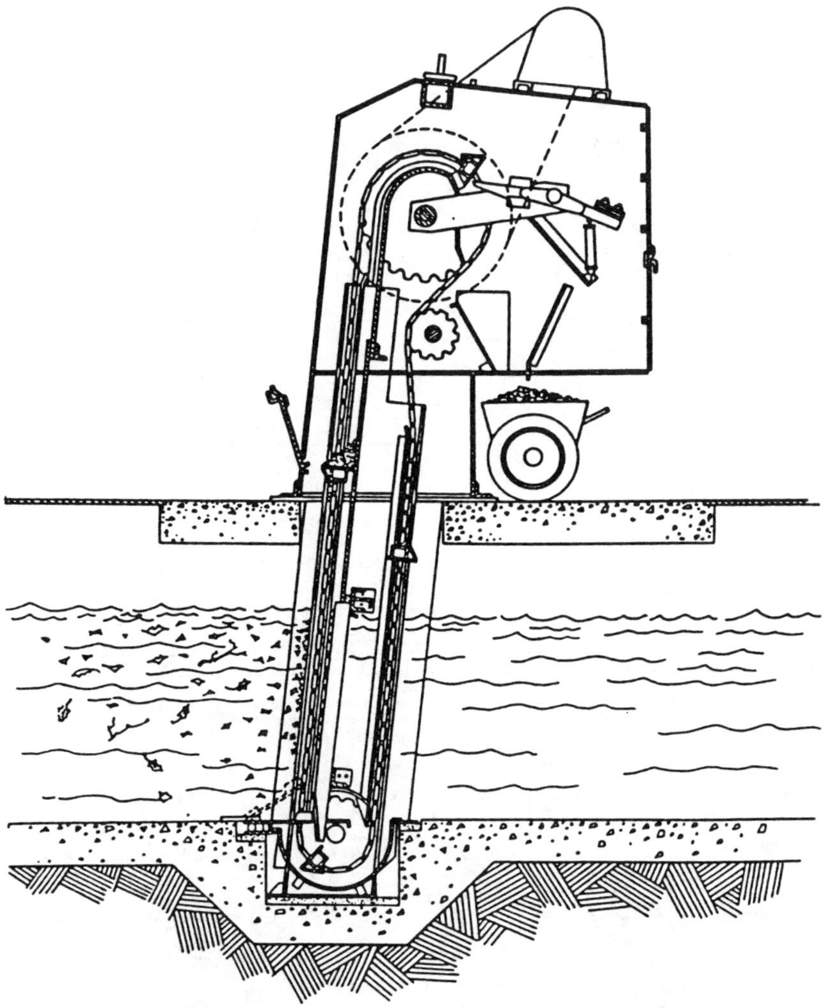

Figure 8.3 Chain- or cable-driven mechanically cleaned bar screen.

and operating mechanisms with the back- or through-cleaned screening configuration are protected by the bars from damage due to large objects in the wastewater stream. The back-cleaning action pushes solids away from the bars, preventing solids from being compressed and jammed between the bars. However, longer rake teeth protruding through the bars are more susceptible to bending and breakage than are the teeth of front-cleaned screens.

RECIPROCATING RAKE SCREENS. The reciprocating rake screen, also known as a "climber screen," can be equipped with a front clean/front return mechanism (Figure 8.4) that minimizes solids carry-over or with a back-clean/back-return mechanism. The up-and-down reciprocating motion of the rake, similar to that of a person raking a manual bar screen, minimizes the

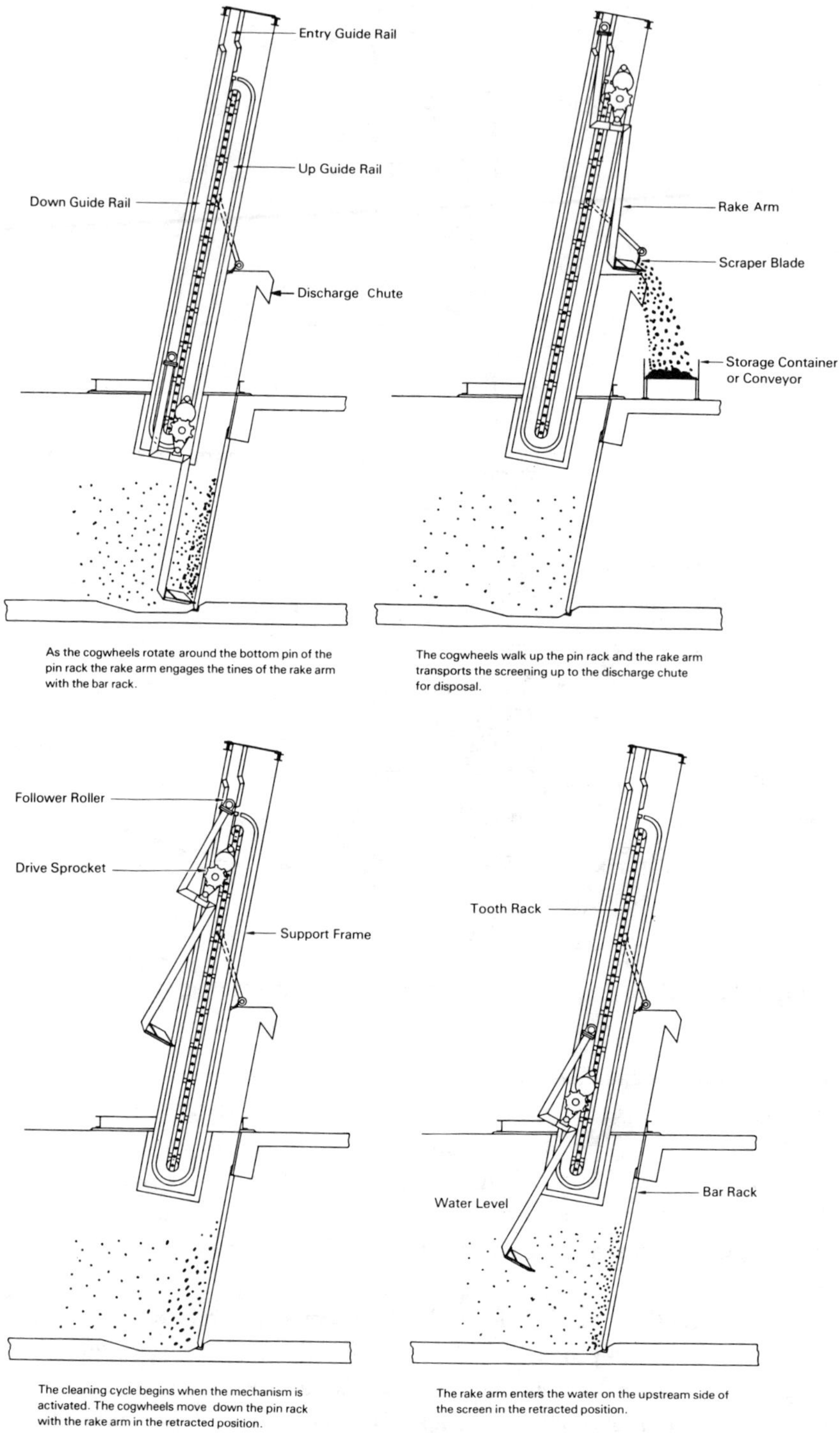

Figure 8.4 Mechanically cleaned reciprocating rake bar screen.

possibility of jamming. These screens are popular because their lack of submerged moving parts allows easy, trouble-free inspection and maintenance without dewatering the channel.

Headroom requirements for reciprocating rake screens are greater than those for other types of screens. The headroom requirement can be determined by adding the vertical depth of the screen to the screening discharge height above the floor.

Although many drive mechanisms are available (chain and cable, hydraulic, and screw operated), the most successful designs include a cogwheel-type drive. For these designs, the entire cleaning rake assembly, including the gear motor, is carriage mounted on cog wheels that travel on a fixed pin or gear rack. The drive mechanism is typically designed to allow the rake to ride over obstructions encountered during the cleaning stroke. In the unlikely event that the rake becomes completely jammed, a limit switch is activated to turn off the drive motor.

A disadvantage of reciprocating screens is the single rake, which limits the reciprocating screen's capacity to handle extreme loads, especially for deep applications where cycle times are long. Also, these systems require high overhead clearance, which might limit their use in retrofits.

ROTATING RAKE ARM SCREENS. The rotating rake arm (or rotary arc) screen (Figure 8.5) features a curved bar rack design to match the path of the rake. This design is particularly applicable to smaller flow/channel arrangements. The rake and arm are attached to a drive shaft and rotate around the centerline of the shaft. As the rake sweeps the front of the bar rack, it picks up

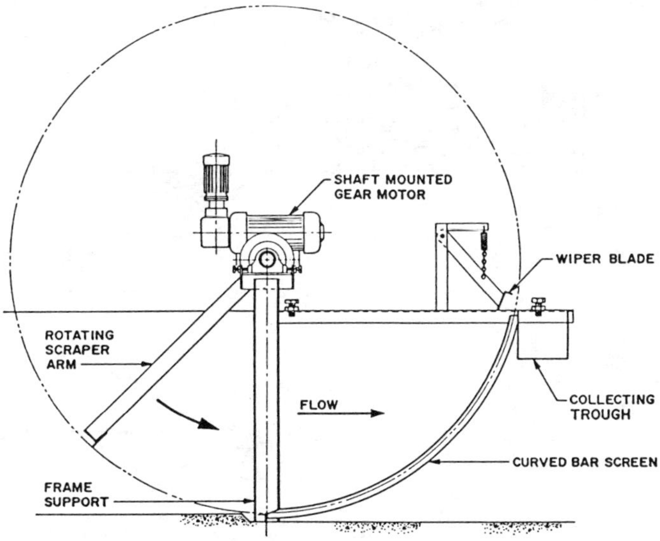

Figure 8.5 Mechanically cleaned rotating rake arm bar screen. Courtesy of Infilco Degremont Inc.

and removes debris that has accumulated. When the rake rises to the top of the bar rack, a wiper blade dislodges debris to a perforated collection trough. Similar to the reciprocating rake screen, the simplicity and lack of submerged moving parts provide ease of inspection and maintenance. Attention must be given to the lateral and vertical space requirements of the screen and the removal of screenings from the collection trough.

Semirotary arc screens (Figure 8.6) are manufactured using a pivoting linkage. As the rake completes its upward cleaning stroke, it disengages and returns to the bottom of the curved bar rack, reengages, and repeats the cycle. Its primary advantage is that it requires approximately 50% of the channel length and headroom required by a full rotary arc screen (Pankratz, 1988).

CATENARY SCREENS. The cleaning mechanism of a catenary screen (Figure 8.7) consists of heavy tooth rakes held against the screen by the weight of its chain. The screen's name, "catenary," stems from the catenary loop formed by the operating chain ahead of the screen. A curved transition piece at the base of the screen allows for efficient removal of solids captured at the bottom. Like reciprocating rake screens, all sprockets, shafts, and bearings are located out of the flow stream to reduce wear and corrosion and reduce required maintenance. Because the cleaning rake is held against the bars primarily by the weight of the chains, the rake can be pulled over large rags or solids that may be tightly held against the screen.

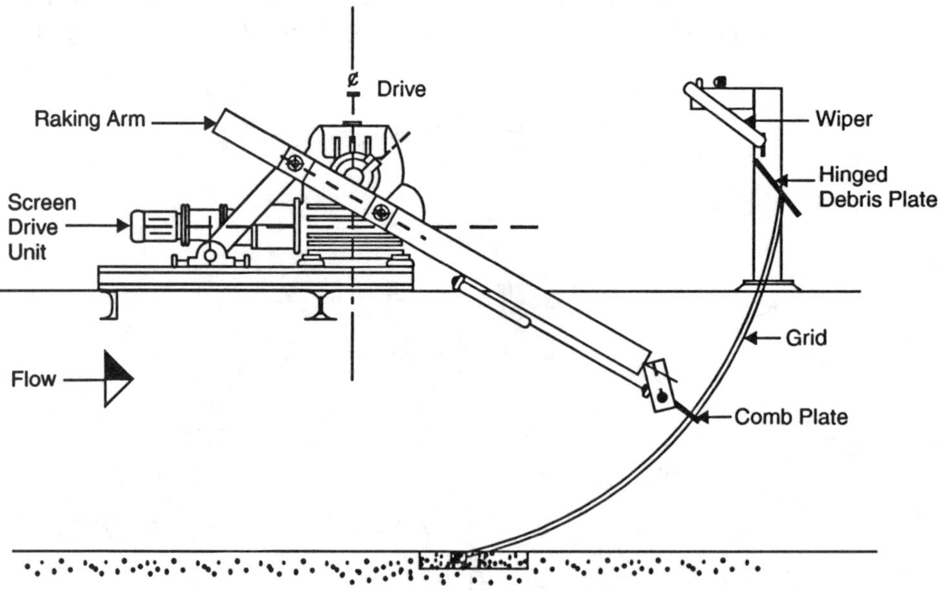

Figure 8.6 Semirotary arc screen. Pankratz, T. (1988) *Screening Equipment Handbook for Industrial and Municipal Water and Wastewater Treatment.* Technomic Publishing Company, Lancaster, Pa. Reprinted by permission of the publisher.

Figure 8.7 Catenary bar screen.

CONTINUOUS SELF-CLEANING SCREENS. Continuous self-cleaning screens (Figure 8.8) consist of a continuous "belt" of plastic or stainless steel elements that is pulled through the wastewater to clean the entire length of the screen. Screen openings are designed with both horizontal and vertical limiting dimensions; the vertical spacing is slightly larger than the horizontal spacing. Continuous screens may have openings as small as 1 mm (0.039 in.), ranging up to more than 75 mm (3 in.). The greater solids-handling capacity of these screens allows smaller openings to be used with greater capture of solids from the waste stream. Also, the continuous screening action of these units allows efficient removal of large quantities of solids. Continuous screens have either a lower gear sprocket or a guide rail at the channel bottom to support the screen elements. Construction of a recessed notch or step in the channel at the screen bottom is a good practice that can help prevent buildup of grit and debris ahead of the unit. Continuous screens are designed to be pivoted up and out of the channel for maintenance. Some screen systems include spray bars and brushes to improve cleaning of screen elements. A disadvantage of continuous screens is possible solids carry-over resulting from the front-clean/back-return design. Another disadvantage is the removal of a high

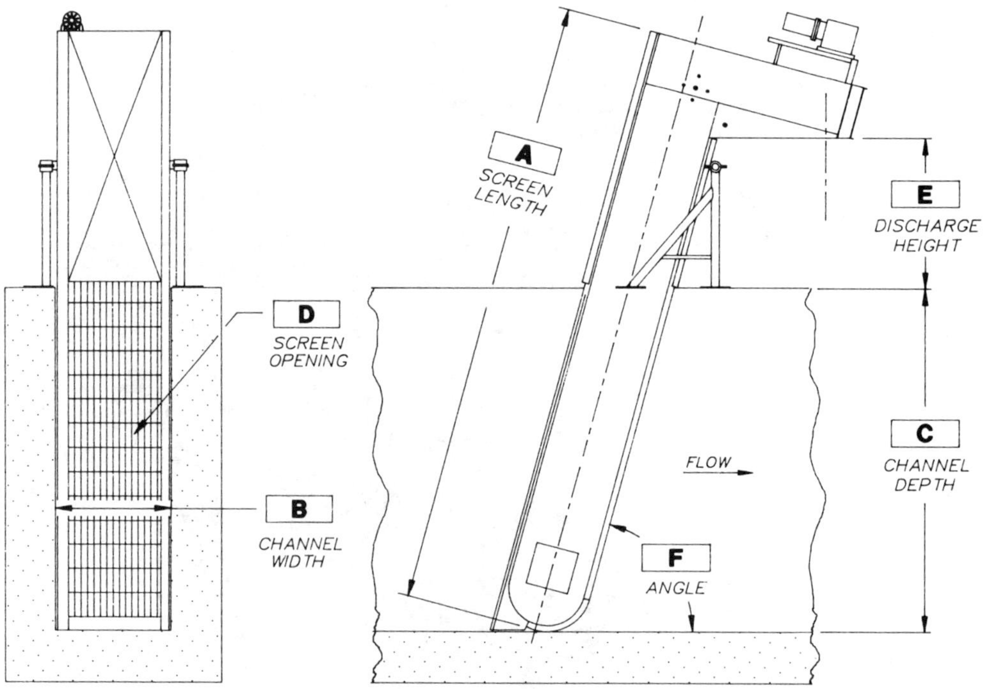

A = 8' to 25'	C = 2' to 16'	E = 3' to 5'
B = 12" to 120"	D = 3/16" to 3/4"	F = 60 to 80 degrees

Figure 8.8 Continuous self-cleaning screen (in. × 25.4 = mm; ft × 0.304 8 = m).

percentage of organic material, which makes storage and disposal of screenings more difficult.

COMMINUTORS. Comminuting devices are installed in the wastewater flow channel to screen and shred material into sizes from 5 to 20 mm (0.25 to 0.75 in.) without removing the particles from the flow. The best application of comminutors is probably in small outlying pumping stations on wastewater collection systems where problems with large solids not readily cut and shredded are usually infrequent. The use of such a device is intended to reduce odors, flies, and unsightliness often associated with screenings and essentially eliminate the operational requirements of screenings removal and handling. Unfortunately, solids from comminutors have caused downstream problems at treatment facilities. Pulverized synthetic materials will not decompose in the sludge digestion process and, if allowed to remain in the sludge, they could bar public acceptance of the sludge for reuse as a soil amendment. Comminutors may create "ropes" or "balls" of material (particularly rags) that can clog mechanical aerators, mixers, pump impellers, sludge pipelines, and heat exchangers at treatment facilities. Because operational experience with comminutors has been generally unsatisfactory, they are avoided in new designs and are being removed from many existing facilities.

SCREENINGS. Both the quantities and characteristics of screenings to be collected and removed require careful consideration before design of screening systems.

Quantities of Screenings. The quantity of coarse screenings to be removed can vary significantly, depending on the bar screen opening, flow characteristics of the municipality, type of collection system, and type of screen. These quantities are difficult to estimate without actual operating data. Data should be obtained whenever possible.

For separate sanitary sewer systems, the most important variable affecting screenings quantity is usually the clear opening between bars. Figure 8.9 represents data obtained from 133 installations of manually and mechanically cleaned bar screens from a general cross section of separate collection systems in the U.S. Average and maximum volumes of screenings per unit of wastewater flow for various bar openings, such as that shown in Figure 8.9, should be adjusted by the design engineer according to the type of collection system and community characteristics.

For bar openings between 25 and 50 mm (1 and 2 in.), the volume of screenings removed per unit of flow is proportional to the clear opening dimension. In this range, for each 13-mm (0.5-in.) reduction of clear opening size, the volume of screenings will approximately double. For screen openings less than 25 mm (1 in.), the volume of screenings removed will increase

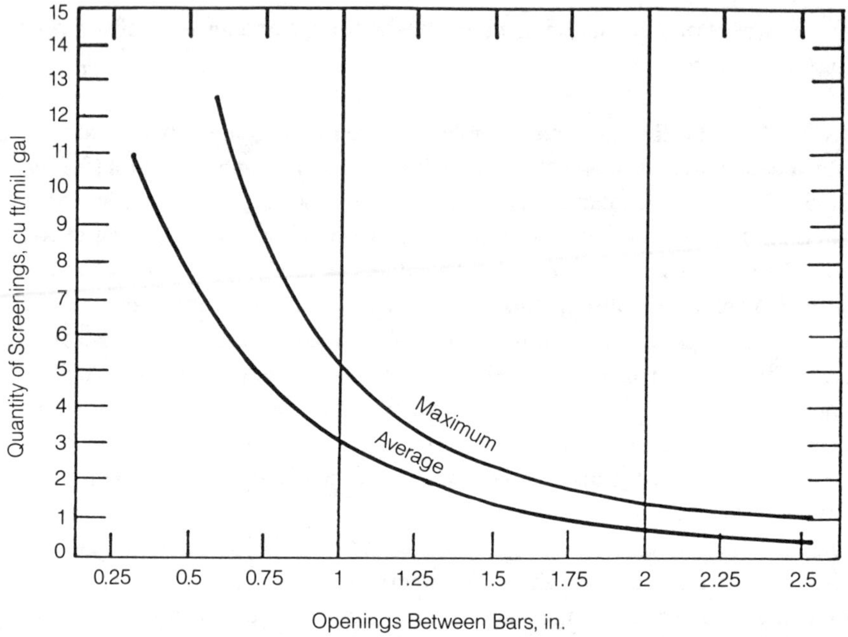

Figure 8.9 Average and maximum volumes of coarse screenings as a function of openings between bars (cu ft/mil. gal × 0.007 5 = L/m^3; in. × 25.4 = mm).

more rapidly as the clear opening decreases to approximately 6 mm (0.25 in.). Removal by trash racks has little relation to the volume of flow.

The quantity of screenings removed will partly depend on the length and slope of the collection system and whether pump stations exist. Screenings quantities may be greater with a short, gently sloping collection system with low turbulence than with lengthy, steep interceptor systems or systems with pumping stations. This difference stems from disintegration of solids exposed to long-duration turbulence. The impact of turbulence will be more significant for smaller bar openings of less than 12 mm (0.5 in.), which capture more organic solids.

Because stormwater and combined sewer systems produce several times the coarse screenings produced by separate sanitary sewer systems, data presented in Figure 8.9 do not apply to the former systems. In addition, the quantity of screenings removed in stormwater and combined systems during wet periods will greatly exceed that during dry periods.

Characteristics of Screenings. Composition and volume affect the disposal of screenings. Coarse screenings consist of branches, bottles, cans, rags, sticks, leaves, food particles, bones, plastics, bottle caps, and rocks. With smaller openings near 6 mm (0.25 in.), cigarette butts, fecal matter, and other organic matter are included in wastewater screenings. The normally large organic portion of screenings readily decomposes; plastics and inorganics are generally not biodegradable. Some of the material in wastewater screenings is inherently odorous and other material absorbs odorous compounds from the wastewater. Some of the pathogenic organisms from the wastewater will also likely enter the screenings.

Screenings, typically containing approximately 10 to 20% dry solids, will have a bulk density ranging from 640 to 1 100 kg/m^3 (40 to 70 lb/cu ft).

Handling of Screenings. The method of cleaning manual or mechanical screens relates to the method of removing and transporting screenings to a disposal site. When designing a manually cleaned screen, relatively small screenings channels [<2 m (<6.6 ft)] are needed, with a drainage plate to allow drainage of the screenings before shoveling. The method used to carry the screenings to a truck or other transport may range from a wheel barrow to a bin carried by an overhead crane or monorail to a conveyor belt system. A belt conveyor, when used, is typically interlocked to start when the screening device starts and run for several minutes after the screening device stops. The conveyor would typically be used to discharge noncompacted screenings to a truck loadable bin.

Compaction of the screenings removed is an option for dewatering and reducing the volume of screenings. This is used primarily for medium to large stations. Such devices, including hydraulic ram and screw compactors, receive screenings directly from the screens and discharge the compacted screenings "sausages" a short distance to portable bins or conveyors. Compactors can reduce the water content of the screenings by up to 50%, with a volume reduction of 6.5 to 1. However, sticks and large objects in the screenings may cause mechanical breakdowns. As a good practice, controls should sense jams, automatically reverse the mechanism, and activate an alarm when the motor overloads. Compactors handle well the increased quantity of solids removed by the smaller openings of continuous screens.

Grinders may be used to chop removed solids and thereby reduce their volume. An automatic grinder can be provided with the mechanically cleaned bar screen to remove solid materials from the wastewater, grind them, and return them to the wastewater, upstream of the screen so as to be rescreened to protect pumps from clogging. The best engineering practice is to dispose of solids once they are removed from the stream. The problems associated with returning ground screenings to wastewater were described previously in the section on comminutors. Ground screenings are easily pumped to a convenient point for dewatering, usually by static screens. The high maintenance associated with grinding, pumping, and dewatering usually eliminates grinding as an economically viable option unless the screenings are to be incinerated. For ultimate disposal in a sanitary landfill, grinding of screenings is typically unwarranted. Should there be any constraints (space, financial, or structural electrical) in the pumping station that prevent the installation of a grinder(s), then pumps with chopping capability (chopper pumps) are recommended.

To complete the handling of screenings, the material must be transported to a disposal site in accordance with local, state, and federal regulations. Lime is sometimes added to the screenings to control odors and insects. Any transportation or disposal alternative must be sensitive to traffic, air emission, water quality impacts, and environmental regulations. The Wastewater Treatment Plant Data Survey (WPCF, 1989b) shows that most plants dispose of screenings at landfills, while a few bury or incinerate the screenings on site. Disposal of screenings as part of refuse or sludge-composting operations is another option.

DESIGN PRACTICE. In selecting screening devices, the following should be considered: when agitated, rags that are comminuted tend to agglomerate into a rag ball that looks like an oily waste rag. This is not a problem at a remote pumping station where only pump protection is required. However, at a treatment plant, the rag ball could disrupt downstream treatment processes.

In bar racks, the smaller the opening between bars, the greater the quantity of screenings that will be removed. Therefore, bar spacing should be small enough to protect the pump but as large as feasible to minimize the amount of screenings that must be removed. If the only criterion is to protect the pump, it is suggested that the distance between the bars be one-third the size of the maximum solid that the pump can discharge.

In a small pumping station, the construction cost of fixed bar screen or comminutor installation may represent a relatively large percentage of the total station cost. Many times, the only solution that can be justified economically is to install a basket screen at the wet well inlet. Basket screens should have sufficient capacity to hold screenings normally accumulated in 24 hours. A convenient raising and lowering arrangement is essential to allow the screenings to be removed for disposal.

Installation. Screening devices should be installed in a positively ventilated area in accordance with safety requirements and guidelines. In addition, devices should be in a well-lighted area that is accessible to operating personnel. Water under pressure should be available for cleaning. Labor-saving equipment for handling screenings also is important. The entire screening installation should be designed to reduce odor release to the outside atmosphere.

Both manually and mechanically cleaned screen installations and comminutor screening device applications should include overflow weir bypass channels or equivalent arrangements that function automatically to limit head differential on the screening device and prevent channel flooding in case of screen-cleaning failure or excess hydraulic flow. Multiple screening devices are desirable in most installations to permit satisfactory screening over the full flow range to which the station is subjected. As flow to the station increases beyond the capacity of one screening device, a second device should be put in service—automatically, if possible. Provisions should be made for isolating screening units for maintenance.

The question of whether screening equipment should be located in an enclosed structure depends on two conditions: the equipment design and the climate. In climates with freezing temperatures, a heated enclosure is necessary. A structure will not only protect the equipment but will also ease maintenance and improve aesthetics. Regardless of whether the screening equipment will be housed, the drive mechanisms of a mechanically cleaned bar screen should be enclosed. In windy areas, the screen rake area and the discharge chute area need covers to prevent screenings from blowing away.

Design. The first consideration in selecting a bar screen for a specific application is determining the maximum normal daily flow and maximum possible storm flow. In connection with flow, the following factors govern the selection of bar screens.

The velocity of flow ahead of and through a screen affects its operation. The lower the velocity, the greater the amount of screenings that will be removed. However, the lower the velocity, the greater the amount of solids deposited in the screen channel. Therefore, a compromise is necessary between a velocity that permits deposition and one that will permit 100% removal of material of a certain size. Selection of an average flow velocity of 0.6 m/sec (2 ft/sec) and a maximum flow velocity of 0.9 m/sec (3 ft/sec) for storm flows is common practice. A minimum velocity as near 0.6 m/sec (2 ft/sec) as possible should be maintained at low flows. Where stormwater is to be handled, approach velocities of approximately 0.9 m/sec (3 ft/sec) are desirable to avoid grit deposition at the bottom of the screens. Screens might otherwise become inoperative when most needed during storms.

The velocity distribution in the approach channel is important to good screen operation. A straight channel ahead of the screen ensures good velocity distribution across the screen and maximum effectiveness of the device. Use of other than a straight approach channel has often resulted in large parts of the flow passing through one side of the screen. The channel hydraulic design should ensure that the flow is uniformly distributed.

The second consideration is determining the proper spacing of bars. Bar spacing should be based on the size of solids to be removed. Clear openings in the range of 25 to 150 mm (1 to 5 in.) are common. Clear openings for mechanically cleaned screens may be smaller.

The third consideration in selecting a bar screen is the size of the bars, which directly affects the efficiency of the screen. Bar size is dependent on the width and depth of the screen channel and the condition of the flow. It is common practice to use 8 mm × 50 mm (0.3 in. × 2 in.) bars up to 2 m (6 ft) in length and 10 mm × 65 mm (0.4 in. × 2.5 in.) bars up to 3.5 m (12 ft) in length.

Other considerations are the size of the channels and the number of screens. Duplicate or multiple bar screens can be provided in a lower capacity range in a single station instead of one large bar screen. Duplicate installations are generally desirable to provide a degree of redundancy. If several bar screens are installed, the screen channels should be so arranged that any unit can be bypassed for service or repairs.

Trash racks are now seldom used ahead of coarse screens or grinders, except for combined systems with significant quantities of stormwater and at prisons with heavy screening quantities. Trash rack bar openings generally range from 40 to 75 mm (1.5 to 3 in.) (WPCF, 1989a).

Coarse screen openings selected range from 5 to 25 mm (0.25 to 1 in.); approximately 50% of the respondents favor openings of 20 to 25 mm (0.75 to 1 in.). A 20-mm (0.75-in.) bar spacing is sometimes too large because twigs, ball-point pens, and other plastics can pass sideways through tee bars. Hence, an increasing number of facilities are switching to continuous-type screens with smaller openings in the 5- to 13-mm (0.25- to 0.5-in.) range (WPCF, 1989a).

Additional design considerations for mechanical bar screens include discharge height to accommodate screen conveying equipment, angle of screen, screen cover to block wind and improve aesthetics, construction materials and coatings for overall unit, drive unit service factor, drive motor size and type of enclosure, spare parts, provision of redundant screen or bypass manual screen, and head loss through the unit.

Instrumentation. For mechanically cleaned bar screens, the following types of controls can be used alone or in combination: manual start and stop, automatic start and stop by timer control, high-level switch, and differential head-level-actuated starting switch on the cleaning mechanism. The differential head-level control system may be pneumatic pressure sensing, ultrasonic level sensing, or direct pressure sensing.

Mechanical bar screen installations should include high upstream level and screen start–fail alarms.

A well-designed screening system will interlock auxiliary equipment such as spray wash, conveyors, or screening compactors with the operation of the screen. This will ensure coordination among the process operations.

Hydraulics. Based on an assumption of steady, two-dimensional flow, clean water head loss through a bar screen can be calculated by using the Bernoulli equation. This equation states that the depth of flow plus the velocity head before the screen equals the depth of flow plus the velocity head after the screen plus friction losses. Actual head loss varies with the quantity and nature of the screenings allowed to accumulate between cleanings. Head loss created by a clean screen may be calculated based on the flow and the effective area of the screen openings. The effective area of the screen openings equals the sum of the vertical projections of the screen openings.

Head loss through a clean or partially clogged bar screen can be represented by the following equation:

$$H_L = K(V_1^{\,2} - V_2^{\,2})/2g \tag{1}$$

Where

H_L	=	head loss,
V_1	=	velocity through bar screen,
V_2	=	velocity upstream of bar screen,
g	=	acceleration due to gravity, and
K	=	friction coefficient.

Head loss through only a clean bar screen is indicated by Kirschmer's equation as follows:

$$H_L = \beta(w/b)\ 1.33\ h \sin \phi \tag{2}$$

Where

H_L	=	head loss,
β	=	bar shape factor (Table 8.1),
w	=	maximum cross-sectional width of bars facing upstream,
b	=	minimum clear spacing of bars,
h	=	upstream velocity head, and
ϕ	=	angle of bar screen with horizontal.

Manufacturers'catalogs offer further details and guidelines in the sizing and selection of bar screens.

Table 8.1 Kirshmer's values of β.

Bar type	β
Sharp-edged rectangular	2.42
Rectangular with semicircular upstream face	1.83
Circular	1.79
Rectangular with semicircular upstream and downstream faces	1.67

GRIT REMOVAL. Grit deposition and subsequent removal are potential problems for stormwater and wastewater pumping stations. Stormwater and wastewater grit materials are typically nonputrescible, have a settling velocity greater than that of organic putrescible solids, and consist of discrete particles. Such materials include sand, cinders, rocks, coffee grounds, metal and plastic objects, and other relatively large-sized nonputrescible organic and inorganic substances.

It is standard design practice to not accumulate or remove grit at stormwater and wastewater pumping stations. Channels and wet wells are designed to provide slopes and velocities that will transport grit through the pumping station without accumulation. Where grit deposition is allowed or planned, the design should include provisions for ease of grit removal by the operators.

Where grit removal at a pumping station is desired, the various methods used in preliminary treatment at wastewater treatment facilities are applicable and are described in *Design of Municipal Wastewater Treatment Plants* (WEF, 1992).

ODOR CONTROL

NATURE AND ORIGINS OF ODOR. Odors from septicity or industrial wastes at a pumping station are a nuisance to operations personnel. They may also be a nuisance to nearby residents, and associated pollutants could pose health hazards to operations personnel. In the past, odors commonly associated with collection, pumping, and treatment systems have been somewhat accepted by the public. Today, offgases, whether a health and safety issue or a nuisance condition, are typically not tolerated by the local community. Consequently, the location of the station and age of the wastewater are factors in the consideration of odor control.

Most odor-producing compounds found in domestic wastewater result from anaerobic biologic activity, which consumes organic material, sulfur, and nitrogen found in the wastewater. Organic odors are typically the result of biological activity that decomposes organic matter and forms a variety of malodorous gases, including indoles, skatoles, mercaptans, amines, and others (WPCF, 1990).

Hydrogen Sulfide. Sulfur is present in human excreta, and sulfates are found in most water supplies. Sufficient sulfur is normally available in domestic wastewater in the form of inorganic sulfates and sulfides or organic sulfides (for example, mercaptans, thioethers, and disulfides) for the production of odorous gases or vapors by anaerobic and facultative bacteria.

Hydrogen sulfide (H_2S) is the most commonly known malodorous gas emanating from wastewater collection and treatment facilities. It is highly soluble (2 800 mg/L at 30°C to 5 650 mg/L at 5°C) in normal domestic wastewater. In addition to its rotten-egg odor, H_2S can cause highly corrosive conditions and is an extremely toxic substance in higher concentrations. Walls and crowns of sewers or closed tanks often have droplets of attached spray or condensation on the surface. Water (such as condensed water vapor in the form of dew or droplets) saturated with H_2S as a result of bacterial action forms sulfuric acid, which is corrosive to lead-based paint, concrete, metals, and other materials. The toxicity of H_2S is on the same order of magnitude as hydrogen cyanide (HCN); death may result in a few minutes when one is exposed to an excess H_2S concentration. For a fuller account of H_2S hazards see *Safety and Health in Wastewater Systems* (WPCF, 1983).

At a pH of approximately 9, more than 99% of hydrogen sulfide dissolved in water is in the form of the nonodorous hydrosulfide ion (HS); while at pH 5, only 1% is in the nonodorous form. If the pH is above 8, there is little trouble from H_2S gas. The relationship between pH and H_2S production is shown graphically in Figure 8.10.

Hydrogen sulfide is produced by reduction of the sulfate ion [SO_4^{2-}] by anaerobic sulfate-reducing organisms. These bacteria thrive at low oxidation–reduction potentials (ORPs) of –0.20 to –0.30 V, pHs of approximately 6 to 9, and a temperature of approximately 30°C. Anaerobic slime growths and

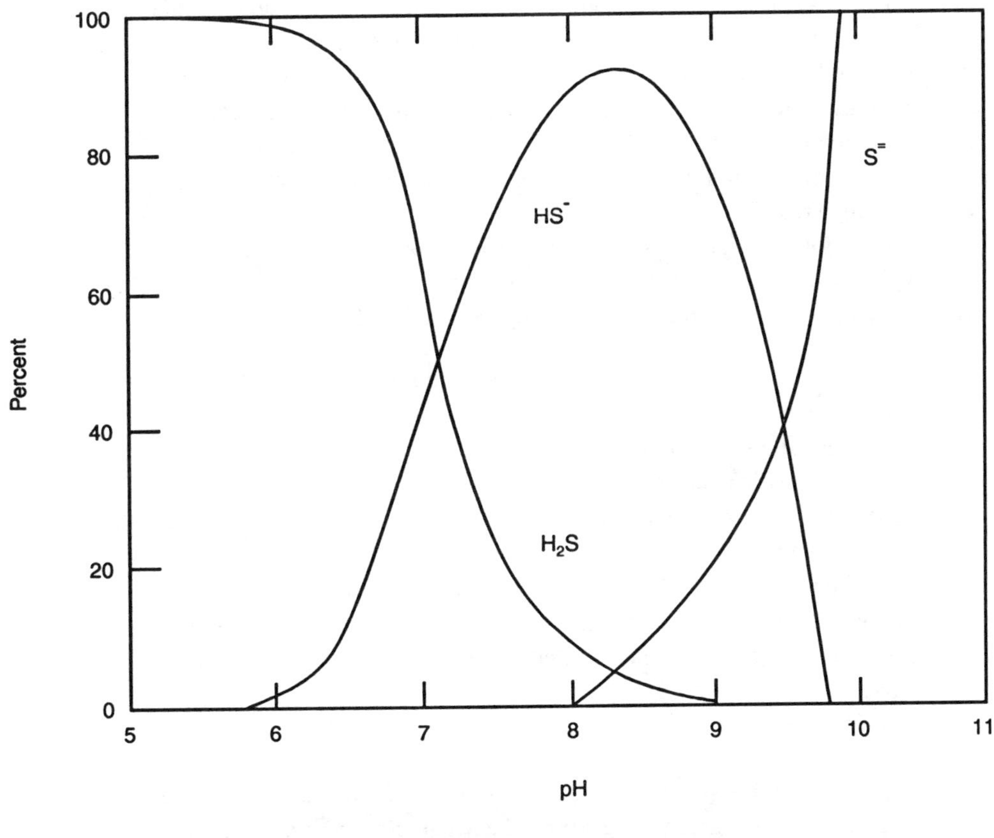

Figure 8.10 Effect of pH on hydrogen sulfide equilibrium. Sawyer, C.N., and McCarty, P.L. (1978) *Chemistry for Environmental Engineering.* McGraw–Hill, Inc., New York, N.Y. Reprinted by permission of the publisher.

sludge deposits that accumulate in sewer lines typically have lower ORPs than wastewater, making them more conducive to H_2S production. Hydrogen sulfide may thus be produced even though wastewater contains up to about 0.3 mg/L of oxygen (O_2). The following section contains a brief description of odor control practices. For more information about odor control, refer to *Odor Control for Wastewater Facilities* (WPCF, 1979).

ODOR CONTROL PRACTICES. Pumping Station Design and Operation. Designing for odor control requires that the engineer assess the effect of the odor on adjacent land use and workers at the facility. Significant criteria are frequency and intensity of occurrence related to the population affected. For example, the effect of an odor on a residential area will be substantially different from the effect on a heavy industrial area where some odor is already present. In certain cases, the cost of providing odor control will be minimal because safety requirements dictate effective venting of units; thus, only odor elimination equipment installations to treat the vented air will be required to process the odorous gas. Because of the nature of odors and the difficulty in preventing their production, every effort should be made in site selection to reduce potential odor pollution. Wind patterns at several locations on the potential site should be studied. Normal wind velocities and durations in each direction should be plotted at each spot to be tested. Sites exposed to winds that continuously dilute any odors emitted are better than sites that are regularly subjected to calm periods (WPCF, 1979).

Accumulations of organic material may produce odors. These accumulations typically result from a lack of regularly scheduled cleaning or basic defects in facility design, such as the following (WPCF, 1979):

- Inadequate velocities in open or closed conduits, allowing solids to separate;
- Square corners that provide pockets where solids can easily accumulate and are difficult to clean;
- Difficult-to-clean rough surfaces on which organic matter and slimes can build up;
- Insufficient nearby hose bibs on pressurized water supply lines;
- Poorly designed drainage systems to carry flushing water cleanly away;
- Inaccessible locations that maintenance personnel cannot easily clean; and
- Inadequate lighting.

Stormwater pumping stations may also be sources of odors if poor maintenance and inadequate housekeeping permit the accumulation of debris, sludge, or trash or the buildup of organic slimes in the wet wells. Regular, careful cleaning is required of stormwater and wastewater pumping stations. The design engineer must provide facilities that permit and facilitate cleaning.

Adequate ventilation prevents obnoxious concentrations of odorants. Vented air should not be recirculated unless odorants have been removed. When untreated odorous air is recirculated even for a short time, odorants may be absorbed in the duct system. Odorous air must be treated or discharged in such a fashion that it will not cause a nuisance either at the facility or to neighbors of the facility.

Whenever odorants accumulate, there is a possibility that a hazardous gas is also present. This makes reliable ventilation doubly important in hazard-susceptible areas. For safety, a flashing light or other similar warning device indicating a lack of air movement should be located at each entrance to a susceptible area; a sign telling what to do if the light is flashing should be posted nearby. This warning device should be triggered by an air movement indicator—not connected to the electric circuit of the fan so as to be triggered by the on–off electric current. Outside fresh air normally should be forced into the ventilated area, pushing the odorous air out through vents at either the top or the bottom of the room, depending on the specific gravity of the gas to be removed. Recently approved National Fire Protection Association's NFPA 820 requires fans on intake and suction. Hydrogen sulfide (H_2S), for example, is heavier than air and would be found near the floor, whereas ammonia (NH_3), which is lighter than air, would be found at the ceiling.

Always investigate the concentration of H_2S at the beginning of the project because a substantial concentration of H_2S (more than approximately 1 mg/L) or a combination of high sewer temperature >22°C (>72°F) and long residence time (>5 hours) in the sewers may

- Be a deciding factor in choosing variable-speed drives;
- Require a sealed wet well or pump sump;
- Alter the location of the pumping station site;
- Require odor control units on the ventilation exhaust system to meet regulatory requirements;
- Increase the capital and operation costs of ventilating system;
- Influence construction details, such as the kind of cement used or the type and lining of the wet well and sewers; and
- Require a specific regimen of washdown operations to keep acid-producing bacteria in check. [Sanks *et al.* (Eds.), 1989].

Odor Prevention. Because most odors generated in and around pumping stations are the result of anaerobic biological decomposition, eliminating or reducing these anaerobic conditions can be an effective way to prevent odors.

FLUSHING/PIGGING. Regular flushing and cleaning can be used to control organic deposits, slime growth, and resultant odor production. If the force main from the pumping station has a long residence time and is a source of odor production, a program of regular swabbing or "pigging" of the force main may be used to clean the line, thereby arresting the slime growth that leads to odor production. An added benefit of improved pumping efficiency may result from force main cleanings. Swabs or pigs tend to be bullet shaped of various materials and outer surfaces to match the specific conditions of the pipe material and type of cleaning to be performed. Pumping stations can be designed or retrofitted to facilitate pig launching. A typical pig-launching arrangement is shown in Figure 8.11.

AIR/OXYGEN INJECTION. Improvement in the oxygen balance can be achieved by the addition of air or oxygen to wastewater. Addition of sufficient dissolved oxygen can prevent or significantly reduce further sulfide generation from occurring and allow biochemical oxidation of dissolved sulfides. Methods of air and oxygen addition include

- Aeration of wet wells,
- Direct injection of compressed air to force mains,
- Use of Venturi aspirators in force mains or lift stations,
- Use of air lift pumps at lift stations,
- U-tube dissolution using either compressed air or Venturi aspirators in either gravity lines or force mains, and
- Pressure tank air dissolution systems for gravity lines. (U.S. EPA, 1985).

CHEMICAL ADDITION. Chemical addition under the right conditions can prevent release of odorous substances. For example, alkalies such as lime can be added to water containing sulfides to raise the pH, thus keeping the sulfides in the ionized form rather than the hydrogen sulfide (H_2S) form. Other chemicals are added to oxidize odorous molecules into more stable, odor-free molecules or raise the oxidation–reduction potential (ORP) and prevent the chemical reduction of sulfates. Chemicals are also added to kill or inactivate bacteria that cause the production of odorous compounds. The use of chemicals at pumping stations should be coordinated with downstream treatment facilities because their operation may be affected by chemical additions. Among the often used chemicals are (WPCF, 1979)

- Oxidizers
  - Ozone
  - Hydrogen peroxide
  - Chlorine
  - Sodium hypochlorite
  - Potassium permanganate

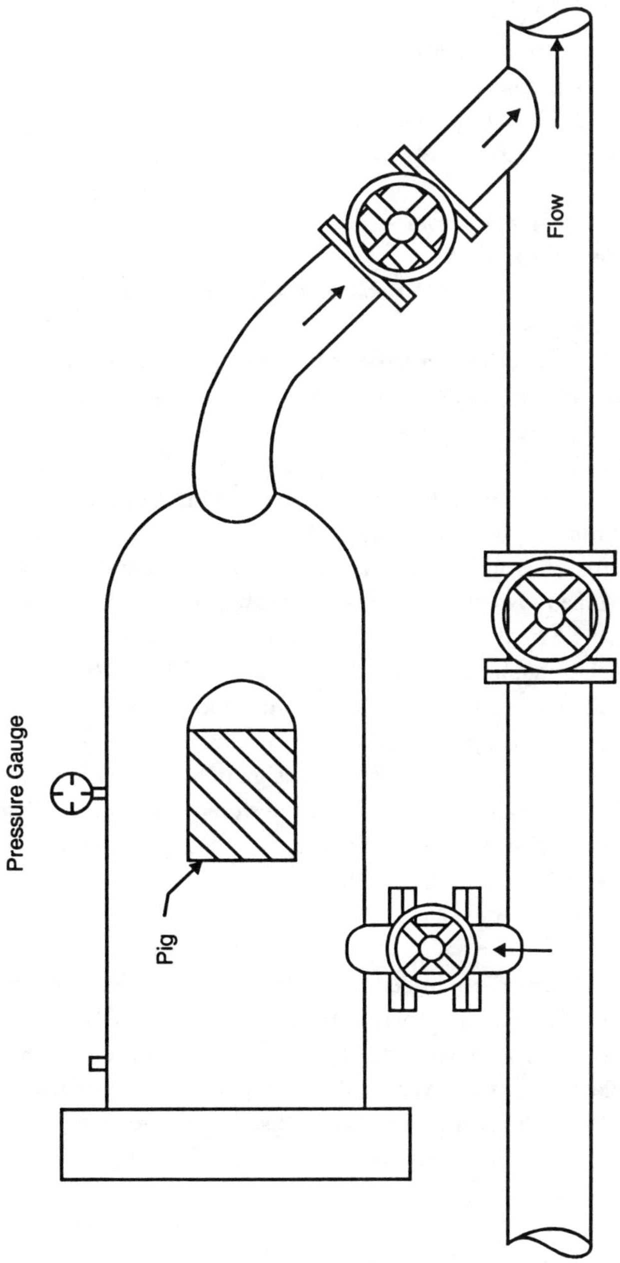

Figure 8.11 Typical pig or swab launcher for force main cleaning.

- Chemicals that raise ORP
  - Oxygen
  - Nitrate
  - Hydrogen peroxide
  - Chlorine
- Bactericides
  - Ozone
  - Chlorine
  - Hydrogen peroxide
  - Potassium permanganate
  - Chlorine dioxide
  - Sodium hypochlorite
- pH modifiers
  - Lime
  - Sodium hydroxide
- Precipitants
  - Trivalent metals such as iron and aluminum

Chlorination. Chlorination represents one of the oldest odor control practices. It is effective in odor control for two basic reasons: it is a strong oxidizing agent and it is an effective bactericide.

Chlorine reacts readily with many compounds in wastewater because it is a strong oxidizing agent. One of the most important of these reactions is with H_2S. Chlorine has been used successfully to destroy sulfides and prevent sulfide from building up in trunk sewers, pumping stations, and force mains. Hypochlorite feed systems are typically used when the requirement for chlorine is less than 2.3 kg/d (5.0 lb/d). If the dosage were 10 mg/L, this would correspond to chlorinating a flow of approximately 230 m^3/d (60 000 gpd). Equipment required for hypochlorite addition includes a hypochlorite storage vessel, metering pump, and diffuser or injector.

Chlorine is a hazardous material. Design of a chlorination station must recognize the need for the safe handling and storage of chlorine. Chlorine is also corrosive to equipment and ventilation requirements are greater. In some jurisdictions, an emergency chlorine gas scrubber system is required where chlorine gas is used.

Hydrogen Peroxide. Hydrogen peroxide (H_2O_2) is a weak disinfectant but a practical oxidant for odor control. Experience has shown that 1.5 to 4.0 mg/L of H_2O_2 is required to oxidize 1 mg/L of H_2S. For prevention of sulfide formation, dosage rates will vary with biochemical oxygen demand, temperature, and hydraulic characteristics of the sewer or force main.

Some of the possible advantages or benefits of the use of hydrogen peroxide for sulfide control are

- It can be used for either gravity sewer or force main applications;
- The chemical feed system is relatively simple and inexpensive;
- Reactions with sulfide or other wastewater constituents produce nontoxic byproducts;
- The decomposition of excess H_2O_2 results in the addition of dissolved oxygen to the stream; and
- With proper dosage, H_2S generation will be suppressed for 3 to 4 hours after H_2O_2 addition.

Adequate safety precautions must be observed in the handling of hydrogen peroxide. Protective clothing, including face shields, must be worn during bulk storage filling or repair or maintenance work where contact or spillage might occur. Any spills should be immediately washed down with water to prevent spontaneous combustion of organic materials in the presence of H_2O_2. Emergency eyewash and shower facilities are necessary for maintenance personnel.

Equipment required for hydrogen peroxide addition is relatively simple, consisting of a storage vessel, metering pumps, appropriate valving and transfer piping, and injection nozzle. For small or intermittent applications, a package dosing system can be used. In this system, hydrogen peroxide is withdrawn directly from containers delivered by the supplier.

Potassium Permanganate. Potassium permanganate ($KMnO_4$) is a strong oxidizing agent and may be used in aqueous solutions at various pH ranges to react with hydrogen sulfide and certain other odor-causing compounds. The actual reaction will depend on specific conditions. Therefore, dosages are case specific and difficult to predict.

Potassium permanganate has been used for sulfide control in collection systems and pumping stations, typically for small or intermittent applications. The relatively high unit chemical cost may be a limiting factor. Some advantages in using $KMnO_4$ are that it is easy to handle, is relatively stable under normal storage conditions, is noncorrosive toward most construction materials, and metering equipment is simple and easily installed.

As with hydrogen peroxide, adequate safety precautions must be observed in the handling of $KMnO_4$. Protective clothing, including face shields, must be worn during bulk storage filling or during repair or maintenance work where contact or spillage might occur. Emergency eyewash and shower facilities are necessary for maintenance personnel.

pH Adjustment. The addition of alkalies to wastewater can be used to adjust pH and effectively decrease the emission of H_2S gas to the atmosphere. The amount of H_2S present depends on the pH of the wastewater and the concentration of sulfides and precipitants.

Aside from the effect that increasing the pH has on the solubility of H_2S, alkalies also serve as good disinfecting agents at high concentrations, thereby inactivating the slime layer that forms along the interior of sewer walls. Sulfide generation in sewers can be stopped for several days if the slime layer is inactivated. If the pH is raised to 12, sulfide generation during the summer months may cease, but may return to 50% of the normal rate in approximately 1 week. If the pH is raised to 13, sulfide generation will practically cease for the first week, but it may return to some degree within 2 weeks. The chemical dosage required to raise the pH to 12 makes odor control by continuous pH adjustment economically unacceptable; however, the infrequency of applying this dosage may prove it to be competitive with other odor control methods in specific situations.

Systems operating successfully today primarily use caustic soda as the alkali. Lime, particularly hydrated lime, may also be used.

Metal Salts. The salts of many metals will react with dissolved sulfide to form metallic sulfide precipitates, thus preventing H_2S release to the atmosphere. For effective removal of dissolved sulfides, the metallic sulfide formed must be highly soluble.

Predominantly iron salts have been used for sulfide control because of economics and the low toxicity associated with iron. Water treatment sludges that are high in iron have been successfully used in controlling H_2S formation and, therefore, may be an attractive economic solution when available.

ODOR TREATMENT. If prevention of odor generation is not feasible or cost effective using the various techniques discussed previously, odorous air within a confined space can be removed and treated before being released to the surrounding atmosphere. Wet scrubbers, activated carbon, and other adsorptive or absorptive processes can be used to remove odor compounds from the air.

Wet Chemical Scrubbing. Wet scrubbing involves contact of odorous gas with a scrubber solution, typically in a countercurrent or crossflow fashion, to allow transfer of the odorants from the gas stream to the scrubber liquid by one or more of the following mechanisms:

- Condensation of odorous vapors,
- Removal of odorous particulates,
- Odor absorption to the scrubbing solution,
- Odor reaction with an oxidizing scrubbing solution, and
- Emulsification of odorous gases in a chemical reagent.

The most common scrubbing medium is water with chemical additives such as buffered potassium permanganate ($KMnO_4$), sodium hypochlorite ($NaOCl$), caustic ($NaOH$), hydrogen peroxide (H_2O_2), chlorine (Cl_2), chlorine dioxide (ClO_2), or ozone (O_3). Potassium permanganate solution is often used as the scrubbing liquid. Sodium hypochlorite, on the other hand, has a faster reaction rate than $KMnO_4$ and is as effective. An important economic consideration when considering $KMnO_4$ is whether the scrubbing liquid will be recirculated, thus obtaining maximum use of the oxidizing chemicals. Recirculation is typically cost effective.

Scrubber water may have to undergo treatment before disposal. If operation is intermittent, accumulated sludge should be drained periodically. Scrubber water disposal is generally not a problem at a wastewater pumping station because the spent scrubber water can be introduced back to the wastewater stream at a rate that will not upset downstream processes.

Activated Carbon Adsorption. Activated carbon adsorption is a commonly used method for treatment of malodorous air. Adsorption is the phenomenon whereby molecules adhere to a surface with which they come in contact. Activated carbon has a high surface-to-volume ratio; thus, a large surface area is available for adsorption in a relatively small volume.

The quantity of material that can be adsorbed in a bed of activated carbon depends on the following factors:

- Concentration of material in the space around the activated carbon,
- Total surface area of the activated carbon,
- Total pore volume,
- Temperature,
- Presence of other competing contaminants,
- Characteristics of the compounds to be absorbed (molecular weight, boiling point, polarity, size, and shape),
- Polarity of the activated carbon,
- Relative humidity of vapor stream, and
- Contact time of vapor stream within the activated carbon bed.

Activated carbon is typically regenerated thermally, although special activated carbon [normally specified for H_2S removal applications, impregnated with caustic ($NaOH$)] can be regenerated chemically using a 50% solution of $NaOH$. Thermal regeneration using multiple-hearth furnaces requires removal of the carbon from the adsorption system.

The design of activated carbon adsorbers involves four fundamental steps:

- Characterize the contaminated air (volumes, concentrations of constituents) and the desired effluent characteristics;
- Select the adsorbent;

- Conduct pilot studies to determine expected performance, useful life of carbon, and design criteria; and
- Apply pilot data to full-scale design.

The importance of full characterization of the contaminated air cannot be overemphasized. Of particular importance is the diurnal variation in air volumes and odor intensity. Failure to accurately estimate these parameters may result in poor performance or higher than anticipated carbon replacement or regeneration frequency. Characterization of specific constituents may also be important in design because special impregnated activated carbon now available may demonstrate improved performance over standard carbon.

OZONE. Ozone is a strong oxidizing agent and disinfectant. Ozone is effective in treating atmospheric H_2S but is not effective in controlling H_2S in liquid. Because ozone is toxic to humans in concentrations as low as 1 mg/L, odorous gases should be drawn into a confined space for treatment. Ozone is an unstable gas and must be generated at the site. Because of the relatively high capital cost associated with ozone generation, consideration of its use typically is only applicable to large pumping stations.

SOIL/COMPOST FILTERS. Soil and compost filter beds have been successfully used to remove odors from relatively small volumes of air containing odorous compounds. The simplicity and economics of soil or compost filters may make them particularly attractive for pumping station odor treatment. Contaminated gas is introduced to a bed of soil or compost through perforated pipe located near the bottom of the bed. As the gas passes upward through the bed, the soil/compost substrate sorbs and oxidizes odorous gases such as H_2S, SO_2, and NH_3. Oxidation products are water, carbon dioxide, sulfate, and nitrate.

Equipment requirements for a typical soil/compost odor filter are minimal. A simple system consists of a perforated pipe network surrounded by gravel to prevent plugging of the perforations and is overlaid with suitable media such as sandy loam or compost. Air piping should be constructed of acid-resistant materials such as polyvinyl chloride (PVC). Valves may be used to direct odorous air to certain portions of the filter bed or bypass parts of the bed to allow maintenance. For moist climates where soil saturation may be a problem, an underdrain system may be used to remove excess water or the beds may be covered to prevent direct infiltration from rainfall. In dry climates, periodic watering may be required to maintain a suitable moisture content in the bed. This may be accomplished manually or through the use of an automated water system.

Design of soil or compost filters depends on the nature and concentration of odorous compounds, air flow rates, target concentrations in exhaust gas, and characteristics of the soil or compost media. Designs vary from simple filter beds with networks of perforated PVC pipe to more elaborate systems with concrete berms, underdrain systems to collect excess water, and automated watering hardware.

Because of the many variables affecting design and performance, it is recommended that pilot studies be conducted before designing full-scale facilities. (U.S. EPA, 1985)

ODOR MODIFICATION, COUNTERACTION, AND MASKING. Odor modification is the name given to a process whereby two substances of given concentration are mixed, with the resultant odor of the mixture being less intense than that of the separate components. In some cases, odor would then be imperceptible.

Odor counteraction is the name given to the phenomenon whereby odor intensity is reduced by adding odor agents that are not chemically reactive to a malodor. Examples of pairs of odiferous gases that, in certain proportions, result in the mixture being odorless or nearly so are ethylmercaptan and eucalyptol, skatole and coumarin, and butyric acid and juniper oil.

Counteraction agents are available to counteract the following odor types:

- Phenolic,
- Aldehyde,
- Amine,
- Solvent,
- Mercaptan,
- Aromatic, and
- Organic, fatty acid.

Odor masking is the name given to a method by which the quality of a malodor is overcome by mixing it with a substance having a strong, pleasant odor. Thus, the pleasant odor effectively masks the malodor, resulting in a less objectionable odor.

Typical masking agents include heliotropin, vanillin, eugenols, benzol acetate, and phenylethyl alcohol. These are organic aromatic compounds that can be effective as short-term or emergency solutions to severe odor problems.

Odor modification, counteraction, and masking applications have met with variable success for wastewater odors. They have often been applied as interim solutions or for short durations during periods of odor emissions. In some cases where odor-masking agents have been used, resulting odors have been considered equally unpleasant. The problem is often further complicated by the subjective human response to odors. In general, prevention of odor emissions and positive control and removal of the odor-causing substance(s) are preferred.

NOISE CONTROL

NOISE CHARACTERISTICS AND SOURCES. Careful consideration must be given to the noise problems in pumping stations. One of the biggest concerns in deciding the location for pumping stations is the effect of noise in residential neighborhoods. Interior noise levels also shall be considered because of the safety of operating and maintenance personnel exposed to long periods of high noise levels.

The most common noise problems in pumping stations come from pumps, electric motors, fans, compressors and blowers, and engine generators. The design engineer should be able to obtain noise data from the manufacturer of any of these pieces of equipment.

APPLICABLE CODES. The design engineer should be familiar with the latest Occupational Safety and Health Administration (OSHA) occupational noise exposure and local environmental control ordinances. The intent of these codes is to protect the worker and the surrounding neighborhood against the effect of high noise levels. See OSHA and local noise ordinances for permissible noise exposures.

If the permissible noise exposure is exceeded in the interior of any facility, hearing protectors shall be used by the worker to avoid permanent hearing loss.

Besides meeting state and local noise ordinances, the design engineer should consider additional noise control measures to prevent any adverse conditions in the community.

NOISE REDUCTION MEASURES. Effective noise control is typically achieved by means of isolation, dissipation, or a combination of the two. This involves the use of silencers, acoustical enclosures, sound barriers, acoustical louvers, and acoustical linings. Additional noise attenuation is achieved by the type of building construction materials used. Manufacturer literature should be used to obtain sound attenuation for this type of equipment and construction materials.

During the design of the pumping station, it is important to know if sound attenuation will be used because additional space may be required to install this equipment.

BUILDING SERVICES

HEATING, VENTILATING, AND COOLING. Heating, ventilating, and air-conditioning (HVAC) systems are required in pumping stations to provide comfort and safety for operating and maintenance personnel, protect equipment, and help control odor emissions. This section of the manual provides the HVAC design engineer with some design criteria, design procedures, and equipment now in use for HVAC systems.

Design Criteria. *CODES AND STANDARDS.* The design engineer should be familiar with applicable state, local, and federal codes such as mechanical, plumbing, fire and electrical codes, U.S. Environmental Protection Agency (EPA) and OSHA regulations, recommended standards for wastewater works ("Ten State Standards"), and the American Society of Heating, Refrigeration, and Air-Conditioning Engineers (ASHRAE) standards.

WET WELLS. Proper positive ventilation design is important in wastewater wet wells because of the possible buildup of combustible and toxic gases (methane and hydrogen sulfide, respectively). Due to the possible presence of these gases, it is desirable to design the wet well so that it is completely isolated from the dry well. Design engineers should become familiar with the minimum ventilation rates required to control these gases to prevent fatal accidents.

The potential buildup of hazardous gases in wet wells of stormwater pumping stations is less likely than in the wet wells of wastewater pumping stations. Therefore, ventilation system criteria for wet wells in stormwater pumping stations are not as stringent.

All ventilation equipment in wastewater wet wells shall be classified as nonspark construction in accordance with the Air Movement and Control Association (AMCA) and all motors and wiring shall be classified as explosion proof per the National Electrical Code (NEC).

According to the "Ten State Standards," the minimum ventilation for wastewater wet wells is 30 air changes per hour intermittently or 12 air changes per hour continuously. Also refer to NFPA 820 and *Design of Municipal Wastewater Treatment Plants* (WEF, 1992) for recommended minimum ventilation rates.

Design criteria for heating loads are based on guidelines found in ASHRAE standards. Wet wells with equipment such as screens shall be maintained at a minimum temperature of 10 to 13°C (50 to 55°F).

DRY WELLS. Heating and ventilating systems in dry wells of pumping stations are designed to remove excess heat from equipment in warm weather and maintain adequate space temperature for maintenance personnel in cold weather.

The heat gain in dry wells is produced by building envelope condition, pump motors, and variable-speed drives. The heat dissipation by the equipment often exceeds the structure heat loss and excess heat must be removed by a ventilation system; in the summer months when the ventilation system cannot maintain temperatures below 40°C (104°F), some type of cooling system may be required.

The heat dissipated by the motors of the pumps should be used in winter to decrease space heating requirements and conserve energy.

As per the "Ten State Standards," the minimum ventilation for dry wells of wastewater pumping stations is 30 air changes per hour intermittently or 6 air changes per hour continuously.

Design criteria for heating and cooling loads are based on guidelines found in ASHRAE standards. Dry well space temperature shall be maintained at a minimum of 13 to 15°C (55 to 60°F) in cold weather and a maximum of 40°C (104°F) in warm weather.

HEATING SYSTEMS. Heating systems typically consist of unit heaters, air handling units with supply fan, heating coils, filters, air dampers and exhaust fans, ductwork, and controls. A boiler (natural gas, oil fired, digester gas, or electric) typically uses an energy source. Self-contained gas fired and electric units are also used as other sources of heat, particularly for small stations.

Unit heaters are sized to take care of heat transmission losses through the building envelope and heating coils are sized to take care of outdoor air (ventilation load). Wastewater wet wells do not require heating unless there is equipment that requires maintenance, for example, screens.

VENTILATION SYSTEMS. Ventilation systems in hazardous areas (wastewater wet wells) shall be designed for 100% positive outside air. These systems typically comprise fresh air intakes (louvers and gravity ventilators), ductwork, supply fans, exhaust fans (roof or in-line), exhaust outlets, and controls.

Class I and Division 1 and 2 hazardous areas shall be monitored for combustible gases or toxic fumes such as methane, gasoline, and hydrogen sulfide. Ventilation rates shall be doubled on detection of 20% lower explosive limit (LEL) for methane or gasoline or 10 mg/L for hydrogen sulfide. If the gas concentration increases, local and remote alarms shall be energized.

In dry wells, the outdoor air ventilation can be adjusted by using mixing dampers to mix return air with outside air, depending on the building heat gain. Efficient ventilation typically is accomplished by forcing air into the bottom of the dry well and exhausting it through the roof. This system of ventilation displaces stagnant air from the dry well and promotes circulation to provide a better environment. Ventilation systems for dry and wet wells shall be entirely separated.

COOLING SYSTEMS. Heat gain in pumping stations typically is greater than the heat loss. If, with the current ventilation system, it is not economically feasible to remove the excess heat, an air-conditioning system may be required. In climates where the outside temperature and humidity are high, a refrigerant cooling system may be necessary; in dry climates, evaporating cooling systems may be more effective.

LIGHTING. Adequate illumination should be provided for all stations, with conveniently located light switches. Shadows and glare should be avoided in the vicinity of machinery and floor openings. Deep wells should have multilevel light fixtures. All fixtures should be so located that maintenance will not be a major problem. It is necessary to have adequate lighting so that maintenance work may be performed in pumping stations. Because small power tools frequently are used for maintenance, electrical receptacles or a portable compressor for air-operated tools should be placed near mechanical equipment.

Design. The following lighting levels are recommended: 540 to 1 080 lx [50 to 100 foot candles (ft–cd)] for offices, 320 to 800 lx (30 to 75 ft–cd) for control areas, 160 to 540 lx (15 to 50 ft–cd) for equipment areas, and 11 to 55 lx (1 to 5 ft–cd) for outdoor areas. Many states now have lighting limitations based on watts per square foot. This limitation should be investigated before the start of the design.

Manufacturer's catalogs provide data on wattage, voltage, lumens, and coefficient of utilization. Using this data, room dimensions; wall, ceiling, and floor reflective percentages; and the number of lighting fixtures can be calculated for each area.

Each room or area should have a National Electrical Manufacturers Association (NEMA) classification rating associated with it. This rating will determine the type of lighting fixture to install. The manufacturer will indicate in the catalog for what type of environment the fixture is suited. Wiring associated with the installation of each fixture shall be in accordance with the National Electrical Code and all applicable state and local codes.

Each fixture should have a high power factor and high efficiency ballast to reduce electrical energy costs. Automatic lighting control should be considered in areas that are periodically visited. Night lighting should also be considered for areas that are not operator attended at night but are periodically checked by guards or other personnel.

WATER SUPPLY. Water frequently is needed in pumping stations for sink and toilet use, cleaning of the building, and maintaining landscaping. Wet wells need to be flushed and cleaned frequently to minimize grease buildup on walls. If wet well walls are not cleaned regularly, grease can break loose, be drawn into pumps, and clog impellers and check valves.

A potable water supply is desirable to all pumping stations. A positive separation (reduce pressure principle device) should be maintained between the potable system and any lines subject to contamination. Warning signs should be posted at all water taps not directly connected to the potable supply. Hose bibbs are desirable at pump pits and screen rooms. If planting around the station exterior requires water, a freeze-proof hydrant for connecting a hose will be useful. Many pumps require seal water to flush the pump shaft and cool the packing within a stuffing box.

Clear water should be provided for sealing stuffing boxes of volute-type pumps wherever possible. Because a connection between the pump stuffing boxes and a potable water supply is a possible source of contamination, some means of preventing backflow of wastewater to the potable water supply is required. Wastewater may be contained if there is a positive break with an air gap between the potable water supply and the waterseal system. Backflow devices, consisting of a system of check and relief valves that protect against backflow, also may be used. This latter method, however, is not approved by some regulatory agencies.

In waterseal systems with an air gap, potable water typically is discharged to a constant-level tank that is located above any possible flood level. Some public agencies require that the potable water line terminate not less than 150 mm (6 in.) above the top of the tank. Discharge of potable water to the constant-level tank typically is controlled by a float valve. The required gland seal pressure may be maintained by locating the tank high enough in the structure to create the required head by gravity or by repumping from the tank.

If a pumped system is used, the waterseal pumps may discharge either directly to the waterseal piping or to a pneumatic pressure tank. In direct pumping systems, a pressure-relief valve customarily controls maximum system pressure and returns excess water to the sump tank. Direct pumpage systems either are designed to run continuously or are electrically interlocked to run when any wastewater pump operates. With continuous pumping, energy usage is high. This system is typically not recommended. For systems discharging to a pneumatic pressure tank, seal-pump operation typically is controlled from the tank pressure. Pump operation will be intermittent and recirculation losses will be eliminated with this system, which also reduces power consumption. Water usage is possibly reduced.

Where conserving water is desirable or where graphite-impregnated pump packing is used, solenoid valves interlocked with pump starters may be installed in individual waterseal lines so that water is supplied to a unit only when the pump is running. However, pumps having a flooded suction with the seal subject to internal pressure when the pump is not operating should be provided with continuous seal water supply.

A supply of approximately 0.03 to 0.06 L/s (0.5 to 1 gpm) for small mechanical and packing seals and 0.12 to 0.2 L/s (2 to 3 gpm) for large mechanical and packing seals has been found to be adequate for estimating waterseal supply requirements. Generally accepted practice has been to provide waterseal pressure at seals at not less than 34 to 103 kPa (5 to 15 psi) greater than the pump shut-off head. Consult pump manufacturers for recommended waterseal supply and pressure. Some pumps have a seal water system that is self contained. They take discharge water and pipe it back to the stuffing box.

DRAINAGE. Interior. Drainage sumps should be provided in dry wells for collection of water for pump packing or seal leakage, condensation or dehumidification, waste, piping leakage, ground-water leakage, and wastewater leakage—either inadvertent or from dismantling of pumps or piping.

Pump stuffing boxes typically are connected to drain by 13-mm (0.5-in.) pipes. Air-relief valves or air discharges from self-priming pumps should have piping appropriately sized to a drainage gutter or sump.

To ensure adequate building drainage, a minimum floor slope of 1 cm/m (0.12 in./ft) should be provided. The station also should have sufficient floor drains in both size and quantity. These drains should be located at the lowest spots or in places where spent water may easily be collected.

Dry-pit drainage sumps should be equipped with automatic level actuated duplex sump pumps. Discharge piping is typically directed to the wet well and must extend above the maximum flood level, with provision for prevention of backflow by siphon-breaker or double check valves to prevent accidental flooding of the dry pit.

UTILITIES. It is necessary to contact each utility early in the design to obtain individuals' requirements and the name of the representative assigned. All telephone conversations should be backed up with confirmation letters. A site visit with each utility representative is recommended.

Electric Utility. Provide the utility with the following information: a load summary of the pumping station, both connected and normally operating; an aerial photographic survey or site plan to mark the location of the pumping station; and the preferred point of service. Discuss with the representative the starting methods and frequency of starts for the proposed pumping station and the location for the point of service.

Obtain the following information from the utility company: a complete current rate schedule applicable to the pumping station, the available short circuit current from the utility power line, transformer specifications, size, impedance, insulation type (dry or oil), primary fuse size, and their policy on underground service. After analyzing rate structures, the demand charges, power factor clauses, and underground service installation, discuss with the representative the advantages and disadvantages of each. Discuss transformer ownership and the utility's maintenance policy if the utility is providing the transformer. All underground service cables must be encased in concrete. Determine whether the costs for the underground service will be shared by the utility company or if the owner is responsible.

Provide the following construction information to the electric company: specifications for any special switchboards, substations, and grounding requirements; site plans; building power plans; single-line diagrams; substation plans; and elevations for service entrance equipment. Also provide the approximate date of service and the bid opening date.

Telephone Utility. Obtain an agreement with the telephone utility on the following: approximate date of service; service point; routing of telephone company cables; the location of its equipment in the owners facility; a clear definition of items and interface furnished by the telephone utility and the owner; telephone terminal facilities (for example, plywood panel and enclosure); telephone extension conduit size and type; telephone extension outlet box size, type, location, and mounting height; and equipment interface requirements (for example, telemetry and surge suppressors).

Provide the following construction information to the telephone utility: specifications for the telephone terminal cabinet, site and building plans containing the telephone requirements, the approximate date of service, and the bid opening date.

Gas Company. The following information should be obtained from the gas company: gas meter location, gas pressures, and size of meter. Provide the utility with location of gas piping for connection to main gas pipe.

OTHER APPURTENANCES

HOISTS. Hoists, either manually or electrically operated (explosion proof in areas classified Class I, Division 1 and 2) and either fixed or portable, should be provided in all stations for the handling of equipment and materials that cannot be lifted readily or removed from the station by manual labor. Moreover, it has been found that hoisting facilities usually pay for themselves in the handling of equipment and materials in the original installation in large stations. It may be advisable to provide overhead bridge cranes and electrically operated hoists in large stations or where frequent use is anticipated. Hoists should be provided over basket screens, submersible pumps, other pumps, or motors and in other locations where it is necessary to lift pieces of equipment or containers that cannot be handled by two persons. Providing heavy lifting "eyes" in the ceiling over pumps and heavy valves and equipment will ease substantially the close handling of such equipment.

To avoid accidents and injuries common to the use of ladders or rungs, personnel elevators are desirable in deep stations. Elevators typically are regulated or certified by local, state, or federal ordinances or laws. Elevators in unattended pumping stations should be equipped with an alarm system connected to a location that is always attended.

Every piece of equipment installed in a pumping station can be expected to be removed during the life of the station. When equipment, such as pumps or motors, must be raised to the ground floor or the loading dock floor, removable grating placed immediately above the pump on each floor through which equipment must travel greatly eases equipment removal. Overhead beams centered above equipment to be removed and equipped with chain hoists or electric hoists are equally important in facilitating equipment removal. Once equipment has been raised to the level of the ground floor or loading dock floor, it is useful to have a beam or monorail leading through double doors to where the equipment can be loaded onto a truck from a loading dock.

Overhead hooks, tripods, and A-frames are useful in disassembling equipment, but they are inadequate for equipment removal. Pivoting cranes and monorails are convenient accessories for the installation and removal of equipment.

As previously mentioned, it is important to provide sufficient room within the station for equipment repairs to be performed and parts to be removed.

Planning also must include easy means of removing installation equipment from the pumping station. For small stations, A-frames and dollies usually are sufficient. For large stations, however, a monorail or a traveling crane may be needed to remove heavy items.

SAFETY FEATURES. Accidental spills of toxic or flammable liquids—such as paint thinner from an industry or gasoline spills or gasoline tank overflows that enter sanitary or combined sewers—may lead to the release of explosive vapors in sewers and pumping station wet wells. Natural gas main leaks often cause explosive gas to enter sewers through pipe cracks or joints. Gas may follow the flow to wet wells and cause explosion hazards. Because dry wells are supposed to be separated from wet wells, the danger that explosive vapors may enter a dry well is reduced; nevertheless, potential danger still exists because of possible leakage around pumps.

Mechanical ventilating systems with ducts extended to within 0.6 m (2 ft) of the bottom floor level in dry wells are effective in minimizing the danger of explosions in these confined spaces. (Explosion-proof motors and controls are discussed in Chapter 6.) For additional information on ventilation systems for wet and dry wells, refer to *Design of Municipal Wastewater Treatment Plants* (WEF, 1992).

Confined-space entry and safe working procedures are discussed in the manual, *Safety and Health in Wastewater Systems* (WPCF, 1983). Safety equipment should be specified from the beginning by the design engineer so that it is included as part of the project's construction costs. If it is assumed that this equipment will be bought later from operating revenues, the safety program that should be initiated immediately at the beginning of the operation.

As noted earlier, considering the frequent necessity to enter pumping stations, it is in the interest of safety for the design engineer to use stairs in place of vertical ladders. Falls from ladders and other devices are the second largest cause of compensation injuries. Ladders that are not permanently anchored are inherently hazardous. However, permanently anchored ladders in wet wells are susceptible to corrosion and should not be used.

Even though confined-space safety procedures require standby workers to be present outside a confined space when entry workers are inside, telephones and radio systems should be installed permanently in buildings or trucks so that help may be used in an emergency. Pumping station emergencies usually require prompt attention.

The following safety features should be included in all stations:

1. Railings should be provided around all manholes and openings where covers may be left off during operation and at other locations where there are differences in levels or where there is danger that the operator may fall.
2. Guards should be placed on and around all mechanical equipment where the operator might come in contact with (or clothing might become entangled with) belt drives, gears, chain drives, rotating shafting, or other moving parts of equipment.
3. Warning signs should be placed near dangerous machinery, hidden obstacles, and other hazardous locations. Good safety practice rules should be posted in appropriate locations.
4. Rubber mats of a suitable type should be provided in front of all electrical equipment where there is any hazard from electric shock. Electrical equipment and wiring should be insulated and grounded properly. Switches and controls should be of the nonsparking type. Adequate lighting should be provided in all locations, especially where there is moving equipment or openings in the floors. Wiring and devices in hazardous areas should be explosion proof. All receptacles should be of the three-wire grounding type outlet, and a ground test of each should be conducted after installation. Portable power tools, extension cords, and trouble lights (including their outlets) should be of the three-wire grounding type. Lighting toggle switches should be "T" rated. When sump pumps are used with a flexible plug-in power cord, the cord and attachment cap should be of the three-wire grounding type. The attachment cap should be locked to the outlet with a locking device and the outlet should be protected by a combination toggle switch with built-in overload protection (fractional horsepower manual starter). It is good practice to provide a prominent sign on the major switchgear stating the service voltage. Disconnect switches for equipment located out of sight of the main panels, or motor control centers should be provided such that the operator may lock out energizing equipment when performing maintenance.

5. Stairways should be used instead of ladders or manhole steps, with straight-run stairways preferable to the circular ones. Stair treads should be of the nonslip type. Vertical ladders should be provided with safety cages. Vertical ladders higher than 2.4 m (8 ft) should be provided with intermediate platforms or landings. Service elevators are desirable in larger stations with relatively deep pump pits. The use of manhole steps should be restricted to areas of infrequent use and, if at all possible, to level changes not exceeding 2.4 m (8 ft). Grab bars should be provided approximately 1.2 m (4 ft) above the level of the landing adjacent to the top step.
6. The possibility of slipping on oil or biological growths that may develop near leaks presents a considerable safety hazard. Thus, proper drainage is important.
7. A telephone is desirable because it will permit the operator to maintain regular contact with the main office. Also, in case of injury, fire, or equipment difficulty, it will enable the operator to obtain assistance as rapidly as possible.
8. Testing and monitoring equipment should be available for checking the gaseous content of the atmosphere (particularly in the wet well), oxygen levels, and toxics. Explosive mixtures also may be encountered because it is almost impossible to eliminate completely the discharge of flammable substances to sewers in large systems. Mechanical ventilating systems in both wet and dry wells will reduce the danger of explosions in confined spaces.
9. Suitable fire extinguishers should be located strategically throughout the station. The use of carbon tetrachloride types should be avoided because of the toxicity of this liquid and its vapor.
10. Safety harnesses with ropes, first-aid supplies, lifesaver floating rings, hard hats, and oxygen respirators should be either stored at the pump station or furnished to each work crew. In the design and construction of pumping stations, all safety requirements must follow prescribed local, state, and national safety laws and codes.

REFERENCES

Pankratz, T. (1988) *Screening Equipment Handbook for Industrial and Municipal Water and Wastewater Treatment.* Technomic Publishing Company, Lancaster, Pa.

Sanks, R.L., *et al.* (Eds.) (1989) *Pumping Station Design.* Butterworth Publishers, Stoneham, Mass.

Sawyer, C.N., and McCarty, P.L. (1978) *Chemistry for Environmental Engineering.* McGraw–Hill, Inc., New York, N.Y.

Water Environment Federation (1992) *Design of Municipal Wastewater Treatment Plants.* Manual of Practice No. 8, Alexandria, Va.; Manual of Engineering Practice No. 76, Am. Soc. Civ. Eng., New York, N.Y.

Water Pollution Control Federation (1979) *Odor Control for Wastewater Facilities.* Manual of Practice No. 22, Washington, D.C.

Water Pollution Control Federation (1983) *Safety and Health in Wastewater Systems.* Manual of Practice No. 1, Washington, D.C.

Water Pollution Control Federation (1989a) Design Practice Survey. Alexandria, Va.

Water Pollution Control Federation (1989b) Wastewater Treatment Plant Data Survey. Alexandria, Va.

Water Pollution Control Federation (1990) *Operation of Municipal Wastewater Treatment Plants.* Manual of Practice No. 11, Alexandria, Va.

U.S. EPA (1985) *Design Manual Odor and Corrosion Control in Sanitary Sewerage Systems and Treatment Plants.* EPA/625/1-85/018, Office Res. Dev., Cincinnati, Ohio.

Index

A

B

C

L

M

N

O

P

R

S

T

U

V

W